EDHCC

00012735

LIBRARY

Weather and Climate

This book is to be returned on or before
the last date stamped below.

LIBREX

Nelson

Contents

LIBRARY

ID No: 00012735
Dewey No: 551.5.
Date Acq: AH94-3

Chapter 1
The atmosphere

It is essential first to look at the composition of the Earth's atmosphere and the behaviour of its layers of gases and vapour in response to the continuous input of radiant energy from the sun. Energy is involved in every physical process that governs the behaviour of the weather, the global distribution of climates and the way the climates change. Energy exchanges between surface and air make for stability or instability in the lower atmosphere and thus affect related weather elements such as wind, cloud and precipitation.

Relationships between local weather and climatic characteristics are examined in sequence: first, weather elements and the influence of air stability; second, how weather systems develop across the globe and their effect within the Earth's major climatic regions; and finally, why such climates are ever-changing. It is thus possible to understand atmospheric characteristics and account for regional variations before debating issues of climatic change in general and global warming in particular.

ITS ENERGY

What is the atmosphere?

The air about us is a compressible mixture of gases and vapours retained by the force of gravity. All but 3% of the atmosphere is within 30 km of the surface, about half its mass is below 6 km. When dry the near-surface air is about 78% nitrogen and 21% oxygen, with small quantities of other gases, including carbon dioxide (some 0.03%). In this lower, denser part, the **troposphere**, there are varying amounts of water and concentrations of tiny particles, such as dust, smoke and salt crystals, which can be sufficient to affect the weather.

The temperature of this lower air falls fairly rapidly with altitude up to the **tropopause** at some 10–20 km; above which the temperature increases in the virtually dust-free, cloudless **stratosphere**. The actual altitude of the tropopause varies with latitude and with seasons (p. 94). Such atmospheric 'layering' is created by incoming **solar energy**, which affects the composition, temperature and movements of air at all levels.

Layers within the atmosphere

The sun's energy reaching the outer atmosphere is of short wavelength, mostly between 0.4 and 4.0 μm (millionths of a metre). Almost a tenth is active, highly penetrative ultra-violet

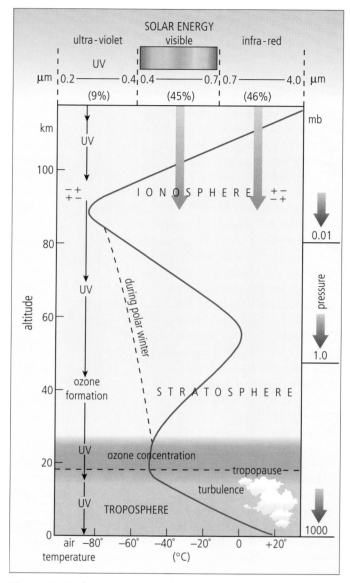

Figure 1.1 A diagrammatic view of atmospheric layering. In reality the altitude of the tropopause varies with latitude: higher near the Equator, lower about the Poles. During winter the air temperature over the polar regions continues to decline to high altitudes.

(UV) energy of 0.2–0.4 μm. Even a hundred kilometres or so up, where air density is very low, UV energy separates electrons from the sparse molecules and raises the temperature. These electrons, with the remaining positive ions, create the ionosphere, which reflects radio waves (Figure 1.1).

The concentration of oxygen molecules increases towards the surface, and from some the UV energy frees oxygen atoms (O), which combine with other molecules (O_2) to form ozone (O_3). This tends to sink and accumulate as an 'ozone layer' in the lower stratosphere. There the sun's energy, partly absorbed by oxygen and ozone, raises the temperature above that of the upper troposphere.

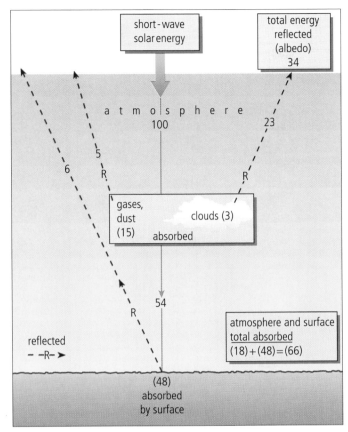

Figure 1.2 The Earth and its outer atmosphere absorb about 66% of the sun's energy received at its outer limits. Most of the remainder is reflected back.

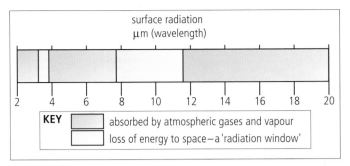

Figure 1.3 The atmosphere absorbs long-wave emissions from the surface, except in the wavelengths indicated.

Warming – the 'greenhouse effect'

About a third of the incoming short-wave solar energy is returned to space by reflection from clouds and the surface, some after scattering from droplets and particles in the troposphere. As the atmosphere as a whole *absorbs* about a fifth, just under a half of that striking the outer atmosphere reaches and heats the surface. This then emits energy of a longer wavelength, which is more readily absorbed by the water vapour, clouds, aerosols and carbon dioxide. These re-radiate heat energy to the air about them and also back to the surface. Such warming by long-wave emissions from a surface heated by short-wave energy is known as the '**greenhouse effect**' – though a greenhouse is enclosed, while freely circulating air exchanges energy in other ways. The lower surface-heated air becomes turbulent, carrying energy up through the troposphere, which also receives latent heat when water vapour condenses to form clouds (p. 11).

Energy gained and lost – a global balance

Figures 1.2 and 1.4 show exchanges in which energy received from the sun and energy lost from the earth and its atmosphere maintain an annual balance. These are only approximations of average conditions. They omit the effects of vegetation and topographical features and mask daily and seasonal variations. In reality a proportion of incoming solar energy is 'stored' for various lengths of time. Energy absorbed by a water surface may cause change of state from ice to liquid to vapour. Then after 'storage' in the vapour it may eventually be released by condensation to droplets (p. 10). Also the sun's energy produces bio-chemical processes in plants and so is stored in vegetation, and perhaps transferred to animals. Some is stored for long periods in fossil fuels, coal and oil, and may be released by human activities.

Nevertheless earth-atmosphere can be seen as a **closed system exchanging energy with space**. Much of the long-wave energy emitted from the surface (mostly about a peak

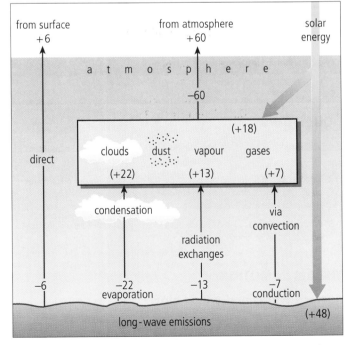

Figure 1.4 Energy absorbed by Earth and its atmosphere is returned to space after numerous exchanges between the surface and atmospheric components.

of 10 μm wavelength) is absorbed and re-radiated by atmospheric gases, water vapour and clouds. A small proportion, notably of wavelength between 8 and 11 μm, can pass through a cloudless sky and escape to space. This 'radiation window' is partially closed by clouds and atmospheric pollutants. Clouds are thus extremely important climatically, for they affect global temperatures in a number of ways – by reflecting incoming solar energy, by absorbing and re-radiating long-wave surface emissions and by closing the 'radiation window'.

Inset 1.1
The flow of energy

- All substances with a temperature above absolute zero (0°K = −273.2°C) emit radiant energy.
- The rise in temperature of a body increases both its radiant output and the proportion of energy of short wavelength.
- Energy not absorbed when it strikes a body is either reflected or transmitted.
- The amount of incoming solar energy scattered by gas molecules is greater in the UV range, and blue light is scattered more readily than red (of longer wavelength) – making for a blue sky.
- Earth emits long-wave radiation over its entire area, but only receives solar energy over the sunlit hemisphere.
- Near the surface water vapour is about 3% of atmospheric volume and is its most important absorber of long-wave energy.

QUESTIONS

1 Why do solar energy emissions peak at about 0.45μm wavelength but those from Earth's surface at 10μm? Consider temperatures.

2 Satellites now monitor the intensity of the sun's energy emissions reaching the outer atmosphere, which are transmitted at a speed of 300 000 km per sec. How long do they take to reach Earth, 150 million km away?

3 Why do the surfaces of mid-latitude deserts experience high daily temperatures yet cool rapidly at night?

THE AIR IN MOTION

Surface winds and winds aloft

As the weight of overlying air can be measured in terms of pressure exerted, meteorologists refer to sea-level as the '1000 mb surface'. Above this the pressure falls – at first by about 1 mb per 100 metres. Higher up the decrease is less rapid as the air becomes less dense; thus the mean height

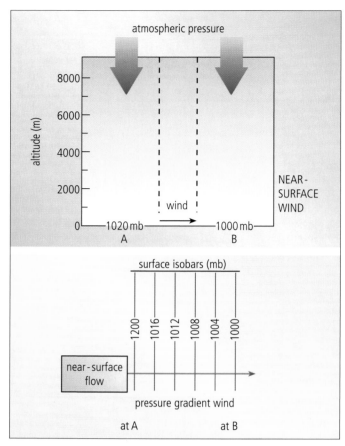

Figure 1.5 *The differences in atmospheric pressure at the surface create a gradient wind.*

of the '500 mb surface' (where the overlying air pressure is 500 mb) is about 5500 m, and that of the 300 mb surface is about 9000 m.

Figure 1.5 shows air flowing in response to pressure differences between places near sea-level, the wind blowing down a pressure gradient represented by isobars. In Figure 1.6 the surface air is calm, but an upper-air wind is established by pressure differences at a particular altitude. Contours of the height of the 500 mb surface are shown, with the upper-air gradient wind. On calm days at the surface cloud movements may indicate air-flows high above.

Surface pressure and winds are often affected by the behaviour of the upper air. As described on pp. 28–32, variations in the flow of fast-moving, meandering upper-air currents may cause air to descend and create high pressure at the surface, or allow air beneath to rise and establish a surface 'low'.

Air circulates about pressure systems

Air flowing over the rotating Earth is apparently deflected – to the right of its line of motion in the northern hemisphere, to the left in the southern hemisphere (**Ferrel's Law**). It is assumed, purely for mathematical convenience, that a force is acting on the gradient wind. This so-called

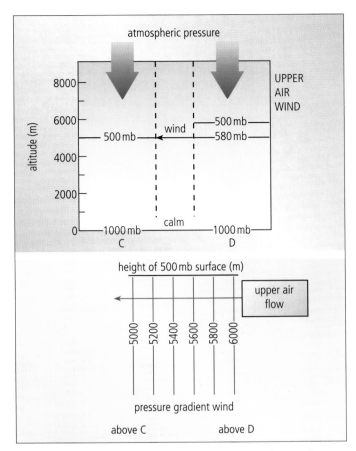

Figure 1.6 *Surface calm, but an upper gradient wind subject to the Coriolis force.*

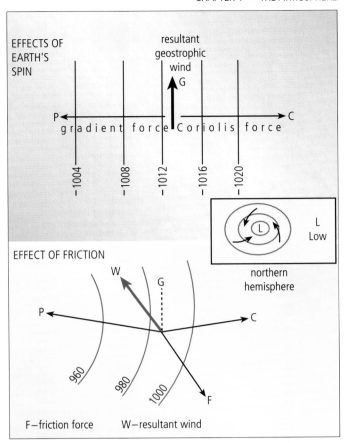

Figure 1.7 *Air flowing towards a centre of low pressure, subject to deflection by the Coriolis force and the effects of surface friction.*

Coriolis force, proportional to the wind speed, is absent at the Equator but increases in strength towards the Poles.

Such a force tends to cause the air to flow as a 'geostrophic wind' *along* the isobars, rather than across them, though its speed is still proportional to the spacing of the isobars. But near the surface **friction** reduces air flow. As the Coriolis force is proportional to wind speed, it is also reduced. Thus there is less deflection of the air as it moves towards the centre of low pressure (Figure 1.7). Over an ocean surface such frictional modification is obviously considerably less than over a land surface.

Air movements away from a high pressure area are generally lighter, but subject to similar modifications.

Surface energy affects air movements

Energy transferred from surface to air creates movements, from local eddies to persistent winds. Thus as heated air expands and rises by **convection**, adjacent air moves in; and as still air chilled by a cold surface contracts, the overlying air sinks, pressure rises and near-surface air moves outward.

Energy exchanges obviously vary with location (e.g. tropical or polar), with season (winter or summer) and with time of day – the effects depending on whether the air is calm or moving over the surface.

The nature of the surface affects its ability to absorb energy and transfer it to the air. Thus for a given amount of energy received water's temperature rises more slowly than that of an equivalent mass of land, and currents may transport energy within it – though warm surface water may overlie cooler, denser water. Hence oceans, with a relatively slow rise of surface temperature, and cooling more slowly than land surfaces, can act as 'heat reservoirs', capable of supplying heat energy to the atmosphere for a long period.

Land does not store heat in this way. Its surface heats and cools quickly. The temperature of a hot desert surface may be some 80°C, and that of air a metre or so above 50°C; yet at night the surface air may be cooled so rapidly that its water vapour condenses to form dew – a potent factor in rock weathering. The contrast in energy storage between land and sea is emphasised in the middle latitudes during winter, when North Atlantic surface temperatures are 10°–15°C compared with those as low as -60°C in continental interiors.

The ability of a surface to absorb energy also depends on its reflectivity (**albedo**), and transparency affects the energy intake and distribution. Figure 1.8 shows how surfaces differ in their reflection and absorption of solar

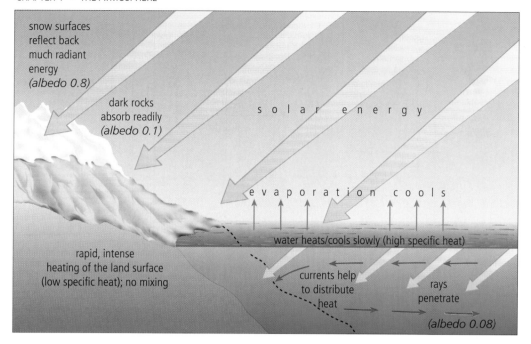

snow surfaces
reflect back
much radiant
energy
(albedo 0.8)

dark rocks
absorb readily
(albedo 0.1)

s o l a r   e n e r g y

e v a p o r a t i o n   c o o l s

water heats/cools slowly (high specific heat)

rapid, intense
heating of the land surface
(low specific heat); no mixing

currents help
to distribute
heat

rays
penetrate

(albedo 0.08)

Figure 1.8 Contrasts in the
ability of surface features to reflect
or absorb solar energy, and water's
ability to distribute energy through
its mass. The albedo is sometimes
expressed as a percentage –
e.g. snow 80%.

energy, their albedo expressed as the fraction reflected.

Even in calm weather differential surface heating of land and sea may create a 'sea-breeze' between cooler denser air over the water and that rising by convection over the land, with the flow reversed aloft where air moves offshore. On a calm, clear night differential cooling can create a 'land-breeze' with a counter flow aloft (Figure 1.9). In the tropics daily sea-breezes may move as far as 100 km inland. On a larger scale inflows of moist oceanic air can become established during summer, especially during the 'wet monsoon' in south-east Asia (p. 37).

Mountain winds

In most latitudes winds arise in mountain valleys under otherwise still conditions. On hot sunny days air over a heated valley floor becomes less dense than the overlying air, causing shallow **anabatic** draughts to move up the valley sides. Their strength varies with local conditions: whether, for instance, a slope is directly heated or in the shade (Figure 1.11). Such updraughts may allow air to move in from adjacent lowland and create an up-valley wind, often with a high level return flow aloft. Under humid conditions clouds form over the valleysides and above the summit.

On cold, still nights upper mountain surfaces rapidly lose energy, so that chilled, dense air flows down slopes as a **katabatic** wind. It may collect above the valley floor or blow down-valley as a mountain wind, with a return current higher up.

Over high, cold surfaces, such as an elevated icesheet, dense air flowing to lower areas creates strong **gravity winds**,

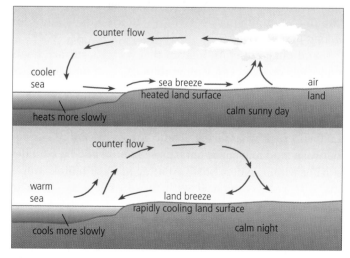

counter flow

cooler
sea

sea breeze

air
land

heated land surface

heats more slowly

calm sunny day

counter flow

warm
sea

land breeze

rapidly cooling land surface

cools more slowly

calm night

Figure 1.9 Sea breezes and land breezes established during hot,
relatively calm weather.

Figure 1.10 Cumulus clouds develop as hot air rises in convectional
updraughts (thermals) over the northern coastland of Tasmania. The sky
remains clear over the cooler waters of the Bass Strait.

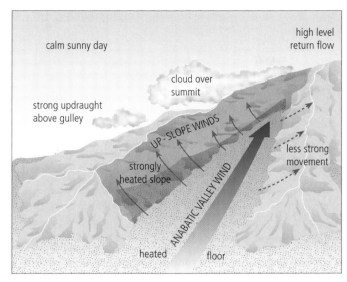

Figure 1.11 *An anabatic valley wind and upslope flows of air rising above the heated valley floor and sunny slopes.*

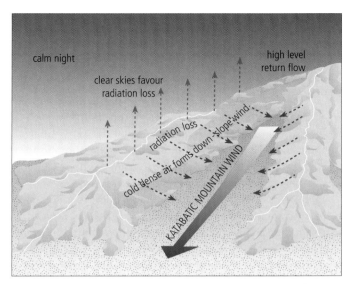

Figure 1.12 *A katabatic wind set in motion by cold dense air sinking from the rapidly cooling valley slopes.*

frequent in Greenland and Antarctica. Under calm conditions mountain glaciers may give rise to similar local winds. In fact mountain climates are very variable, and high surfaces gain or lose energy very rapidly through the thin atmosphere.

Surface obstacles affect airflow

When air passes over a rough surface friction decreases its horizontal speed and eddies are set up; but as the airflow encounters a constriction it increases speed to take it through the narrows. An isolated hill causes the air to accelerate over the upwind side, though friction retards the flow; but beyond, in the lee, as the air spreads and slows there is relatively low pressure, and here eddies develop. In the lee of a mountain range such eddying creates waves, often with clouds at their crests, extending away in parallel lines, as in Figures 1.29 and 1.30.

Strong turbulent winds dry agricultural land and erode top-soil. A tree shelter-belt may reduce air speed downwind for some 10–15 times the height of the barrier. Eddying in the lee may be a problem; though through-flow at trunk level reduces this and extends the protected zone downwind.

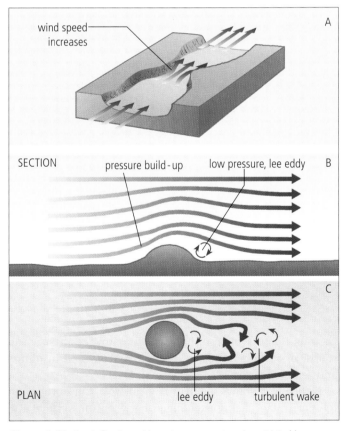

Figure 1.13 A: *air flowing with a given momentum is restricted by a narrow channel and increases its speed – the Venturi effect.* **B** *and* **C**: *An obstacle disturbing the flow causes pressure variations, with eddying beyond.*

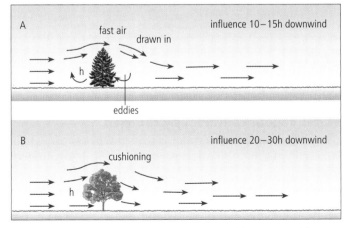

Figure 1.14 A: *A dense wind-break reduces speed but creates eddying.* **B**: *A barrier allowing through-flow partly checks velocity, but extends its influence twice as far downwind.*

Inset 1.2
Regional winds

These are regularly experienced during particular seasons in many parts of the world. Causes include:

- Winter high pressure over central Europe with low pressure in the Mediterranean causes the cold outblowing **bora** to surge through coastal mountains to the Adriatic, with gusts of up to 200 km per hour.
- Depressions – low pressure systems (p. 31) – move eastward through the Mediterranean during spring and early summer drawing dry **sirocco** and **khamsin** winds from the Saharan high pressure zone, bringing dust over North African coastlands and across the sea.
- Such winter depressions may also cause cold stagnant air from high mountains and plateaux of southern France to

be funnelled through the Rhône Valley as the strong, cold **mistral**, affecting the delta area.

- During spring high pressure over southern California's desert interior forces the hot, dry, dusty **Santa Ana** westward through the mountain valleys, often damaging the coastal fruit blossom.
- Low pressure inland of near-coastal mountain ranges draws moist air up and over the heights, causing drier, warming air to descend and move inland – known in the case of the Alps as a **föhn** wind (p. 16), and beyond the Rocky Mountains as a **chinook**.

Locate the regions concerned and turn to pp. 16, 56 and 60 which emphasise the causes of these particular conditions.

QUESTIONS

1 What part does the force of gravity play in (a) retaining Earth's atmosphere and maintaining its structure, (b) surface air pressures and related wind movements?

2 Why is it necessary to regard the Coriolis effect as due to a 'force' and why does it vary with latitude?

3 The so-called 'Ballot's law' states that, for the northern hemisphere: 'Stand with your back to the wind and low pressure will be towards your ......, the high pressure towards your ......'. 'Left' or 'right' – which and why?

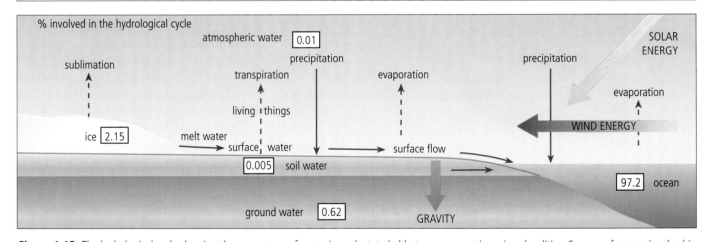

Figure 1.15 *The hydrological cycle showing the percentage of water in each state held at any moment in various localities. Sources of energy involved in transferring water from one part of the world to another are indicated.*

MOISTURE, TEMPERATURE AND AIR STABILITY

Water vapour continuously recycled

Over 97% of the planet's water is held in the oceans, though it is part of a continuous circulation – **the hydrological cycle** – fuelled by solar energy. The atmosphere receives water vapour from the evaporation of liquid water, by sublimation direct from ice to vapour and by transpiration from plants. There the water vapour may condense and remain suspended as droplets in cloud, fog, or mist, or fall as rain, hail or snow; or it may condense as dew on a cold surface – all forms of **precipitation**.

Much of the precipitation finds its way back to the oceans by surface run-off, though water is 'stored' for

various lengths of time beneath the surface in soil and rocks, or in lakes and ice masses, or retained in plants and animals.

Figure 1.15 shows the hydrological cycle, the storage and the percentage being exchanged between the surface and atmosphere. The proportion of water vapour seems small, but its equivalent in rain would cover the Earth's surface to a depth of 25.5 mm. Globally there is roughly a constant amount of water vapour in the atmosphere, but great inequalities of precipitation and evaporation over land surfaces and oceans.

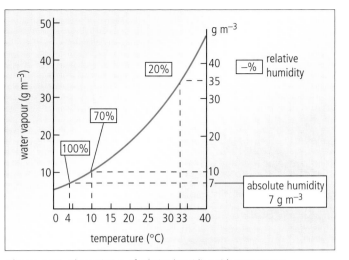

Figure 1.16 The variation of relative humidity with temperature.

The relative humidity of the air

At any place the amount of water vapour in the air varies greatly. In the lower air it tends to be some 2% of the volume, but varies from very little to about 5%. The amount tends to decrease upward, so that the mean vapour content at 1200 m is only a tenth of that at sea-level, though the actual amount is ever varying.

The mass of water vapour in a unit volume of air (g/m³) is the **absolute humidity**. However, meteorologists usually refer to the *mass* of water vapour in a given *mass* of dry air (g/kg) – the **mass mixing ratio** (x).

The water vapour content may also be stated as its contribution to the atmospheric pressure – say 20 mb out of an atmospheric pressure of 1000 mb. Cold, dry air might have a water vapour pressure of less than 2 mb, whereas in moist, tropical air it could exert a pressure of 15–20 mb. At any given temperature there is a limit to the amount of water that can be held as vapour. When the air is saturated any fall in temperature should cause condensation.

The absolute humidity of a volume of air does not change unless water is added or removed, but its relative humidity varies with temperature. **Relative humidity** is the proportion of the *actual* mass of water vapour in a given volume of air to the *maximum* amount that *could* be contained at that temperature; it can be expressed as a percentage. Thus we can find the relative humidity by comparing the actual vapour pressure to that which *would* be exerted if the air were saturated *at that temperature*.

Figure 1.16 shows that air with a water vapour content of 7g/m³ is saturated at 4°C. If the vapour content remains the same, the relative humidity at 33°C is only 20%. The water vapour content that could be held at saturation level at various temperatures is indicated. Thus 'dry' air over a hot desert may contain as much water vapour as saturated air over Arctic waters, or indeed more.

If unsaturated air cools and its relative humidity increases until it is completely saturated, the temperature at which further cooling leads to condensation is the **dew-point**. If the dew-point is below 0°C, some of the moisture condensed may form ice – as snow, white frost, or the tiny crystals of high (cirrus) clouds. In practice, condensation to liquid often occurs at temperatures well below freezing point.

For droplets to form in the atmosphere tiny particles are needed to act as **nuclei of condensation**. These may be microscopically small, such as dust particles or salt crystals, which are common in the lower troposphere. Particles with an affinity for water (hygroscopic substances) are the most effective nuclei of condensation. In the absence of nuclei, cooling may continue far below dew-point without condensation, and the air becomes **super-saturated** with water vapour. Such conditions are more likely in the higher, purer atmosphere, above some 6000 m.

Precipitation in many forms

Heat energy required to evaporate a particular quantity of water is known as **latent heat** in the sense that it remains 'latent' or 'hidden' until that quantity of water vapour condenses to droplets, when the same amount of energy is released to the surroundings. When condensation occurs in rising, cooling air the heat released may then keep that air warmer than the surrounding air, and thus it is more buoyant (Figure 1.20). The condensation at first forms clouds of droplets, usually of radius of 0.001–0.05 mm. Some, by condensation or collision, may become large enough to fall as rain against the updraughts. If the temperature has fallen below freezing point, droplets may form ice crystals, which coalesce as **snow** flakes.

Figure 1.17 The Taranaki peninsula in New Zealand's North Island on a summer's day. Moist surface air of low relative humidity rises up the conical slopes of Mt Egmont, its relative humidity increasing until a saucer-like cloud forms and persists about the summit.

Figure 1.18 *Inversion conditions as air subsides over the western-central Andes, causing a cloud layer to persist above still cool air in the deep valley. Further east, towering clouds form where moist, unstable air from the Amazon basin rises up the steep eastern slopes.*

Hail is usually formed in towering cumulonimbus clouds (p. 15). Strong updraughts carry a droplet high enough for it to freeze. As it falls there is further condensation about it before it is again carried up by air currents. The process may be repeated many times before the resulting hailstone falls to earth, its interior showing concentric shells of ice.

When calm, moist air close to the ground is cooled below dew-point, droplets may remain in suspension as **mist** – or **fog** if visibility is less than a kilometre. Sometimes a valley floor receives cold, dense air from higher ground, so that **inversion** develops, with the cold air below and warmer air above, favouring mist or fog formation from ground-level upward.

Fog may also form in moist air moving slowly over a cold surface. Off the Newfoundland coast, where warm air above the Gulf Stream passes over cool waters of the Labrador Current, there is frequent **advection fog**.

Figure 1.19 *Fog forms near Cape Point, where the warm Aghulas current meets cold waters from the south off this south-western tip of South Africa.*

When the air temperature is well below freezing and the water vapour super-cooled, minute ice crystals may be deposited on grass and tree leaves as **hoar frost**. In very cold, slightly moving, damp air, crystals of **rime** may cover the windward side of trees.

As clouds themselves are masses of tiny suspended droplets or crystals, their form varies a great deal, revealing conditions in the atmosphere about them and indicating high airflows or vertical air currents. Before considering cloud types in detail, it is necessary to appreciate how the temperature and humidity of rising air are inter-related.

Temperature changes as air rises

The rate at which air temperature lapses (decreases) with altitude is known as the **lapse rate**. Close to a heated surface the lapse rate often far exceeds that above it. The first hundred millimetres or so tends to have a micro-climate of its own. Thus above a hot desert surface there can be a fall of 20C° in the first metre, compared with a global mean of 0.6C° per 100 metres. On a sunny afternoon in the tropics a high lapse rate may exist up to several hundred metres above the surface.

Consider a mass of dry air rising, expanding and cooling. (Unsaturated air is described as 'dry' as long as its moisture content does not condense, and so affect the lapse rate.) Such air cools at a **dry adiabatic lapse rate (DALR)** of about 1.0C° per 100 metres.

The term **adiabatic** implies that no heat is transferred between it and the surrounding 'environmental' air, whose own temperature decrease with height is known as the **environmental lapse rate (ELR)**.

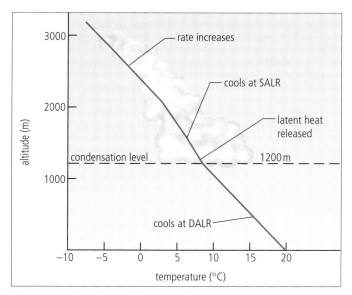

Figure 1.20 *As moist air is forced to rise and cool, condensation releases energy and checks the cooling rate. The SALR curve shows that the rate tends to rise again with altitude.*

If the water vapour in the rising air becomes chilled to dew-point, any condensation releases latent heat energy, so that the air now cools at a slower rate – the **saturated adiabatic lapse rate (SALR)**. At first this is about half the DALR, some 0.5C° per 100 metres, but as the air continues to rise it varies. Higher up, with less moisture and with temperatures between 0°C and −40°C, the rate of cooling may increase to some 0.75C° per 100 metres, and at about 12 000 m the value usually approaches the DALR.

Air induced to rise from the surface may, however, encounter a ceiling of sinking, warming air – a state of **inversion** with, in a sense, a negative lapse rate, as warmer air overlies the cooler. This occurs in the relatively stationary high pressure systems over sub-tropical deserts and over temperate continental interiors in winter. In moist temperate latitudes less permanent inversions occur from time to time, as when cold dense air masses move below a layer of warmer air (p. 31). Advance knowledge of lapse rates, revealing whether approaching air is stable or unstable, is a valuable aid to forecasting.

Stable and unstable air

When air is caused to rise, its behaviour depends on differences between its own properties and those of the surrounding air. Figure 1.20 shows humid air forced to rise, perhaps on encountering a mountainside. Initially cooling at the dry adiabatic rate, it remains colder than the surrounding air; but at 1200 m condensation occurs, releasing heat energy, so that it cools more slowly. Eventually it becomes warmer than the environmental air and more buoyant – an **unstable** condition.

Unstable moist air rising by convection above a hot surface may create towering clouds, surging to great heights (Figure 1.26). Incidentally, apart from temperature factors, air containing water vapour is usually lighter than the same volume of dry air.

When an upward movement is so resisted that it tends to return to its former level, there is atmospheric **stability,** as in Figure 1.21A where air forced to rise cools at the dry adiabatic rate and remains cooler and denser than the surrounding air.

QUESTIONS

1 Why does rising air cool?

2 Why may air rising and cooling at the *dry* adiabatic lapse rate in fact have a considerable vapour content?

3 Why is the saturated adiabatic lapse rate less than the dry lapse rate?

Figure 1.21 *In A and B air is forced to rise. Stable air in A cools at a steady rate, while that in B becomes unstable (as in Figure 1.20). In C cold air remains close to the surface.*

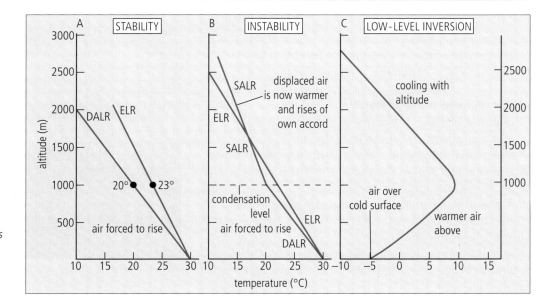

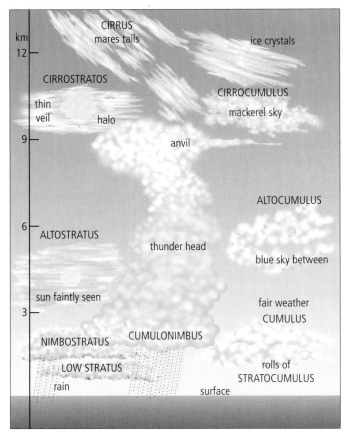

Figure 1.22 *Types of cloud.*

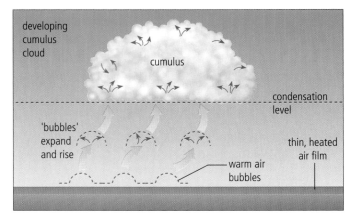

Figure 1.23 *Condensation in individual bubbles of warm, moist air produces a mass of cumulus.*

Figure 1.24 *Billowing masses of cumulus cloud develop in moist air drawn over the central Australian tableland during summer.*

Figure 1.25 *Cumulonimbus developing above the condensation level brings showers to this dry scrub pastureland in central Australia.*

Clouds and atmospheric conditions

Clouds are useful indicators of relative stability, their forms tell a great deal about atmospheric conditions.

Low clouds (from the surface to about 2000 m)

Stratus, a dense, uniform, grey layer of droplets, similar to ground-level fog, can form well above the surface. Ragged and usually shallow, it may drift slowly along. As a thicker layer producing rain, it is known as **nimbostratus**. With clearing weather there is a tendency for the cloud to billow into rolls, as **stratocumulus**, with open sky between.

Clouds billowing to higher levels

Cumulus is formed when convection is strong, with bubbles of warm air (thermals) rising from the heated surface, and at a particular height the water vapour in each condenses. Billowing cloud masses then soar above this condensation level. These may grow in size and develop vertically until they reach more stable conditions at higher levels. Such clouds tend to die away towards evening as convection currents lose strength.

In some clouds the rising air remains unstable to considerable heights, and the summits of towering cumulus may reach some 10 000 m, where super-cooled droplets

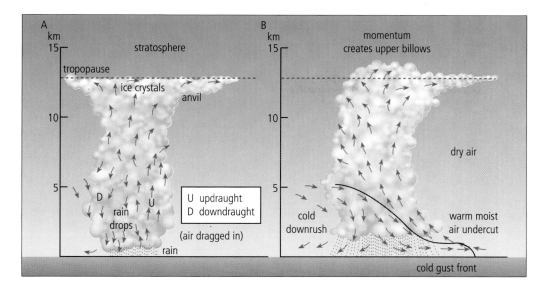

Figure 1.26 As cumulonimbus forms in rapidly rising air, the downflowing air currents countering strong updraughts drag in surrounding air. As in **B**, strong surface wind may lash rain forward against air moving into the system.

and ice crystals form in the higher parts, releasing more energy and increasing the rate of updraught. On reaching an upper surface of stable air, perhaps at the tropopause, the cloud mass with its tiny crystals extends horizontally into an anvil-like form.

Cumulonimbus clouds are apt to produce torrential rain and hail. Lightning and thunder frequently occur as accumulated static electricity, created mainly by updraughts, discharges from cloud to cloud, or cloud to earth. There are also downdraughts, which on reaching the cloud base spread out in cold gusts (Figure 1.26).

Middle clouds (usually formed between 2000 m and 6000 m)

Altostratus is a thick cloud layer with sufficient water to form a darkish cloud, through which the sun may faintly gleam. It frequently forms where moist air moves up a wedge of cold air, as in advance of the warm front of a depression (p. 31), where it may give fairly steady precipitation.

At this height, under fair weather conditions, layers of air of different humidity and density sometimes flow over one another, causing billows of **altocumulus** to form, usually with blue sky between lines of individual clouds.

High clouds (forming generally above 6000 m)

Cirrus clouds, composed of ice crystals, are usually feather-like in appearance. They too form high in advance of a warm front, where warm air rises over a mass of cold air. Their long bands of 'mare's tails' are thus an early suggestion of unsettled weather. Small globular **cirrocumulus** clouds may also form, often in lines, as a 'mackerel sky', giving a similar weather warning. At these heights a thin sheet of ice crystals, **cirrostratus**, may give a milky appearance to a blue sky, with a halo around sun or moon as light is refracted through the crystals.

Figure 1.27 Ragged cloud beneath altostratus about a secondary low encircling a cyclone, which has given Queensland pastures 300 mm of rain. Water pours through the erosion channels.

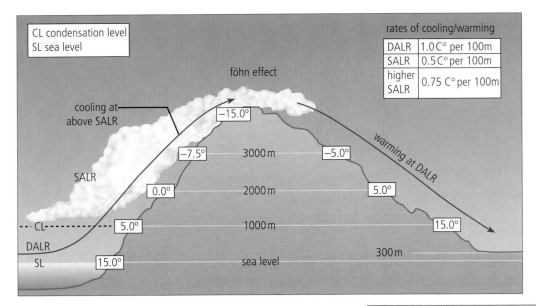

Figure 1.28 An induced airflow across a mountain system. Variation in lapse rates allows considerably warmer, drier air to stream away in the lee, usually with some turbulence.

Up and over high ranges – the föhn effect

Where mountain ranges or scarps deflect wind upward, the amount of precipitation depends on the direction of flow and the moisture content and stability of the air. It also depends on local relief, such as steep slopes, funnels and other topographic features. In the lee of mountains, in a 'rain-shadow' area with drier conditions, the air tends to eddy and form waves that produce rolls of lens-shaped clouds.

As moist but unsaturated air moves up windward slopes, it cools adiabatically by about 1C° per 100 metres; but on continuing to rise above condensation level, and thus gaining latent heat, it cools at the slower saturated adiabatic rate of about 0.5C° per 100 metres, though rather more rapidly higher up. Figure 1.28 shows changes in air temperature at 1000 metre intervals.

As air descends the lee slopes, it gains heat at the dry adiabatic rate and warms relatively rapidly. At equivalent altitudes it has higher temperatures than on ascent, and may blow away from the lee slope 10–20C° warmer than on the windward side, sometimes more. With a higher temperature, and absolute humidity reduced, its relative humidity is very low.

Such föhn winds, as they are known, occur in spring and autumn in the European Alps, when low pressure over

Figure 1.30 Clouds over the foothills in the lee of New Zealand's Southern Alps show effects of air rotation and the development of lee waves.

north-central Europe draws air over the mountains from the south. Their onset, with rapid thawing, is apt to trigger avalanches. Similar winds occur in other mountain systems. The Chinook wind of the Rocky Mountains tends to cause rapid snow-melt in spring, clearing pastures and farmlands east of the ranges.

Cloud observations and atmospheric conditions

It is now obvious that cloud forms and behaviour can provide a great deal of information about environmental air conditions and also about weather in the immediate future. It is often possible to observe whether cumulus is building and perhaps developing into cumulonimbus or whether it is declining. Observing the cloud fringes even for a short time may show visible thickening, indicating continuing instability; or they may appear to thin – the evaporation indicating that conditions are becoming more stable.

High sheet cloud is likely to indicate stability with inversion at a particular altitude, even when, as Figures 1.31 and

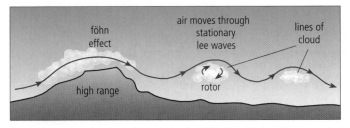

Figure 1.29 Air forming eddies and rolls of cloud in the lee of a mountain range as downdraughts create rotor effects.

Figure 1.31 *Clouds in vigorous updraughts over islands in a Swedish lake flatten beneath altostratus formed under inversion conditions, but pierced in places by air currents.*

Figure 1.32 *A head of billowing cumulus in energetic updraughts, which carry it into clear air.*

1.32 show, there is such turbulence at low levels that surging updraughts pierce the cloud layer.

WEATHER CONDITIONS

Recording the weather

As Chapter 2 considers how particular weather systems develop and how they are indicated on weather maps, it is useful at this stage to look at the symbols shown on such maps and the instruments used to record weather elements on a daily basis in a small weather station.

As shown in Figure 1.33, a rain gauge and Stevenson Screen are set up in favourable positions. With its louvred sides, a metre or so above the surface, the Screen provides shade and allows air to flow over maximum and minimum thermometers and over a wet and dry bulb hygrometer, whose different readings are converted in a set of tables to show air humidity. The Screen should be clear of trees and buildings that could affect air flow.

A mercury or aneroid barometer will indicate air pressure, or, if available, a barograph with pen and revolving drum will give continuous values, as shown in Figure 2.39. A wind vane, erected to operate at least 10 metres from the surface and clear of obstruction, will give wind direction; while a revolving-cup anemometer with a rev-counter will provide wind speed. Alternatively, wind speed can be estimated using the Beaufort scale, whose numbers (0–12) range from 0=calm, 4=moderate breeze, through 6=strong breeze, 9=strong gale, to 12=hurricane force wind.

Cloud cover is estimated in eighths (oktas) (see Inset 1.3), the type of cloud noted and its height estimated. Some local stations measure the duration of sunshine with a Campbell-Stokes recorder, whose glass sphere focuses sunlight onto a sensitive card (Figure 1.33). For educational purposes

such daily readings can build up a comprehensive view of local weather conditions.

Numerous weather stations across the globe, some large, some small, transmit their data to a meteorological office, which combines the information on weather maps (synoptic charts). The main centres also receive continuous information from satellites. A major problem is that large stretches of the Earth's surface, such as oceans and arid areas, provide scanty information from scattered stations – hence the value of even a small station like that in Figure 1.34. The satellites can now transmit information about

Figure 1.33 *A weather station with rain gauge, Stevenson Screen and sunshine recorder.*

Figure 1.34 A small, fenced weather station in the semi-arid Ethiopian highlands, near Lalibela.

conditions over the oceans or remote continental surfaces, but constantly changing energy flows in the lower air often make for abrupt changes in the weather, which surface stations are equipped to pick up.

Inset 1.3 shows the various symbols used on weather charts to represent conditions at a particular station, though the actual information from stations to central office are transmitted in code.

Inset 1.3
Weather symbols

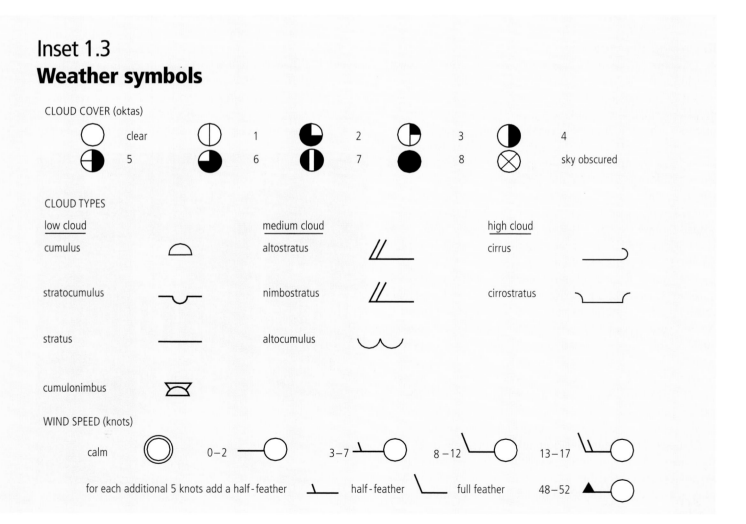

CLOUD COVER (oktas)

clear   1   2   3   4
5   6   7   8   sky obscured

CLOUD TYPES

low cloud		medium cloud		high cloud	
cumulus		altostratus		cirrus	
stratocumulus		nimbostratus		cirrostratus	
stratus		altocumulus			
cumulonimbus					

WIND SPEED (knots)

calm   0–2   3–7   8–12   13–17

for each additional 5 knots add a half-feather   half-feather   full feather   48–52

WEATHER CONDITIONS

mist	═══	snow	❄
fog	═─═	showers (rain)	
drizzle	𝟵	showers (snow)	
rain	●	thunderstorm	

surface air

front of cold air

front of warm air

cold air below/warm above
(occlusion)

Symbols combined as station model

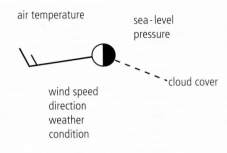

air temperature

sea-level
pressure

wind speed
direction
weather
condition

cloud cover

Station model

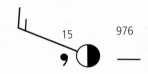

15   976

QUESTIONS

1 Describe precisely the weather conditions shown by symbols and statistics in the Station Model opposite.

2 **a** Give a general explanation for the variations in weather over various parts of Western Europe from information provided by the simplified map for this mid-winter day (Figure 1.35).

   b Explain the influences of the main centres of high/low pressure.

   c How do the symbols for winds and sky cover help to explain the variations in weather in different parts of the British Isles during this day?

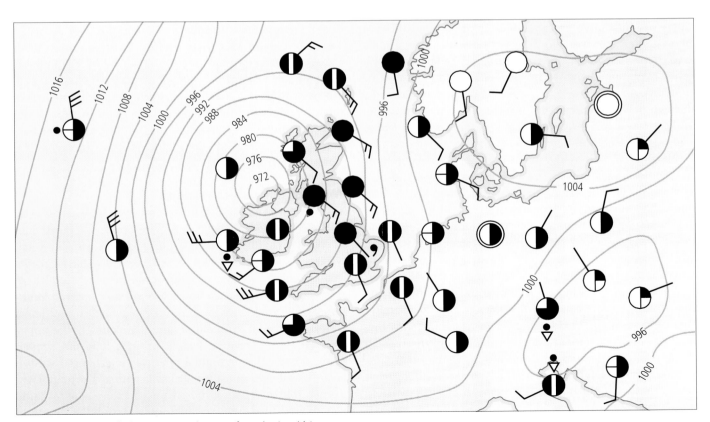

Figure 1.35 *A simplified synoptic weather map for a day in mid-January.*

CLIMATIC INFLUENCES ON PLANTS AND SOILS

Atmospheric conditions naturally affect plants. As Chapter 3 considers classifying climates by the vegetation they support, it is useful to provide an outline here of how weather elements *combine* to affect both the local vegetation and the soils that support plant species.

Plant species and water availability

The temperature and humidity of the air affect the rate at which plants absorb water through their roots, distribute it through their organs and lose it through leaf pores (stomata). Even where precipitation is distributed through the year, cold periods can inhibit growth, or necessitate leaf loss by deciduous species. Where temperatures remain high but seasonal precipitation is followed by lengthy drought, successful plant species are those adapted to conserve water – some by a mechanism that can close their leaf stomata (p. 81). In really arid regions, deep-rooted species may survive, while others become dormant, awaiting a rainy spell – though even here torrential rain may run off the surface and not enter soil-water storage and thus be unavailable to plants. By contrast, with excess surface water the aquatic conditions favour species adapted to receive sufficient aeration – like the water hyacinth, which is sustained by bulbous leaf stalks.

Plant responses to temperature

There are maximum and minimum limits beyond which a plant cannot survive, and between these is a range most favourable for growth. A temperature may be favourable yet other conditions, such as particular light intensity or air humidity, may adversely affect plants. Thus olive trees may survive periods of dry cold, but then a cold damp spell, with no lower temperatures, damages the tissues of leaves and buds.

Plant organs exist under different temperature conditions at various stages of growth, their stems, leaves and root systems developing in different environments. Figure 1.36 shows this for rainforest plants in Java and for plants in sandy soil in East Anglia. Such maximum and minimum figures suggest that a sharp spring frost in eastern England will affect plants according to the height of particularly sensitive organs and whether it is an air frost or a ground frost.

A plant's response to night temperatures may affect its distribution. Many can only flower, and thus perpetuate themselves, when night temperatures are sufficiently low in relation to day temperatures. In the case of the English daisy, if the day temperature reaches 26°C, the night temperature should fall below 10°C. Quite small variations from the necessary night temperature may govern the variety and distribution of plants.

The albedo of a surface (p. 7) is also important, and the

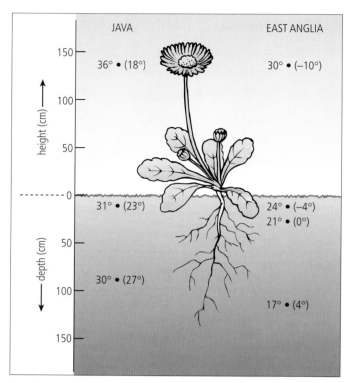

Figure 1.36 *Temperature extremes that various parts of a plant may experience – as measured on farmland sites in Java and in East Anglia.*

colour and the texture of a soil affect its temperature; thus at midday in Bedfordshire thermometers buried 100 mm under loam, loam and soot, and loam and lime recorded 18°C, 22°C, and 15°C respectively. The vegetation itself affects surface temperatures. In southern England in summer, adjacent surfaces of sand and lawn gave temperatures of 55°C and 44°C respectively. Above a hot surface the air temperature decrease with height (the lapse rate) may be many times that of the surrounding air, thus creating turbulence and affecting evaporation rates from the surface and from the plants.

Weather and soil conditions

The combined effects of temperature and soil-water greatly influence the delicate balance between soils and vegetation. In cool wet conditions, acidic water passing through the soil washes down (leaches) many of the bases essential for plants – compounds of sodium, calcium or magnesium – and leaves silica in the now infertile upper layer. Under hot wet conditions, however, soil clays tend to release silica and leave concentrations of iron and aluminium oxides in the upper soil. This may be quite fertile, but if a long dry season follows, it may harden into a relatively infertile red surface layer (laterite). In moist tropical soils with a ready supply of oxygen, bacterial activity is rapid, decomposing organic wastes and providing soluble nutrients for the plants to absorb – a nutrient cycle found in tropical rainforests (p. 52).

Other weather conditions obviously have different effects. Low temperatures and excess water hinder the bacterial decomposition of plant debris, leaving peaty material. While persistent drying tends to draw up salts that may accumulate at, or near, the surface (Figure 3.58). However, soil–plant relationships are complex and, apart from weather elements, the properties of the soil usually respond to material derived from a parent rock. Thus soils on limestones, being basic, tend to resist the leaching of other bases.

Solar energy and plant growth

Light energy required for photosynthesis, which provides plants with carbohydrates, varies from place to place and seasonally, according to latitude. It is also affected by cloud cover and by shading from other plant species. Plants establish themselves to meet their light requirements in relation to higher and lower foliage (Figures 1.37 and 1.38).

The duration of light, or of succeeding darkness, is important. In short-day plant species, flowering is promoted by long nights and short days; while long-day plants flower when the daylight hours are greater than a particular value. For optimum crop yields scientists aim to produce strains related to a certain length of daylight.

Figure 1.37 Epiphytes attached to the upper branches of a rainforest tree, freely receiving insolation and atmospheric elements. Some, like the bromeliads, have water-absorbing scales.

Weather conditions and human comfort

Those other life-forms, people, must adapt to regional weather. We need to maintain a body temperature of 37°C and are comfortable when environmental temperatures are about 20°–25°C. Clothing or shelter help us cope with the cold and we remain active in hot weather mainly by controlling body temperature through sweating, though a combination of heat and humidity makes for heat exhaustion and discomfort. In a hot dry climate it is obviously necessary to drink water and to replace salt lost when sweating.

When a particular temperature is forecast, there may be mention of what we might *feel* the temperature to be, for this relates to humidity and wind speed. Table 1.1A shows how

A	air temperature °C	apparent temperature °C		
	35	34	42	58
	30	29	32	38
	25	23	25	27
	relative humidity %	20	50	80

B	air temperature °C	apparent temperature °C		
	10	9	2	−3
	−4	3	−8	−12
	−12	−14	−32	−38
	wind speed m/sec	2.0	9.0	18.0

Table 1.1

this varies with the proportions of water vapour in the air – the relative humidity. In cool air, strong movements tend to cause 'wind-chill', a sensation felt at various wind speeds that suggest the 'apparent temperature' (Table 1.1B).

Figure 1.38 Plant diversity in the forest of New Zealand's South Island. Broadleaf trees, tree ferns, shrubs and herbs adapt to light energy at various levels; while 'strangler' epiphytes send down roots that branch, multiply and twist about the trees.

SUMMARY

This chapter has emphasised that all weather conditions are related to exchanges of energy within the atmosphere. The prime source of energy is the sun, but the lower atmosphere gains most of its energy from the heated surface by radiation and convection and from the condensation of water vapour. Differences in atmospheric pressure set the lower air in motion, in a direction apparently subject to deflection as it moves over the spinning globe.

The importance of water vapour as an atmospheric component cannot be over-emphasised. It is the most important recipient of energy in the global warming process; it releases energy on condensation; and, on conversion to cloud droplets, it affects surface temperatures, reflects radiant energy and is the source of precipitation.

QUESTIONS

1 Why does local air stability tend to change during a hot summer day? Suggest why the amount of cloud and their shapes tend to vary as evening approaches.

2 What causes updraughts to continue within a convection storm cloud?

3 Which types of cloud may indicate:
   a the likelihood of thunderstorms?
   b a period of steady rain in the near future?
   Describe the atmospheric conditions likely to be involved in each case.

4 Explain various ways in which atmospheric inversion may become established and favour persistent fog? How may human activities increase fog density?

5 Suggest and account for possible advantages for the temperate farmland that lies in the lee of such ranges as the Rocky Mountains or New Zealand's Southern Alps.

6 Why may a mid-latitude desert maintain its general aridity even though the absolute humidity of the air above is high?

BIBLIOGRAPHY AND RECOMMENDED READING

Barry, R. and Chorley, R., 1998, *Atmosphere, Weather and Climate*, Methuen

Hanwell, J., 1989, *Lapse rates, Geography Review*, 2(3), 7 and 2(4), 19

Oliver, H. and Oliver, S., 1993, *The ins and outs of sunshine, Geography Review*, 7(1), 2

Sumner, G., 1996, *The nature of precipitation, Geography*, 81(3), 247

Washington, R., 1996, *Mountains and climate, Geography Review*, 9(4), 2

Wright, D., 1984, *Meteorology*, Blackwell

WEB SITES

BBC Weather Centre –
http://www.bbc.co.uk/weather

Meteorological Office (UK) –
http://www.meto.gov.uk/

The on-line guide to meteorology –
http://www.2010atmos.uiuc.edu/(gh)/guides/mtr/home/rxml

Chapter 2
Weather systems

Seasonal weather conditions obviously vary considerably between one part of the world and another. This chapter describes the development of weather systems characteristic of particular regions, in response both to solar energy inputs at that latitude and to energy transferred from elsewhere by circulating winds and ocean currents. Such circulations themselves are set in motion by seasonal gain or loss of energy by land and ocean surfaces, transferred to affect the temperature and pressure of air in contact.

THE LOWER AIR

Energy received varies with latitude

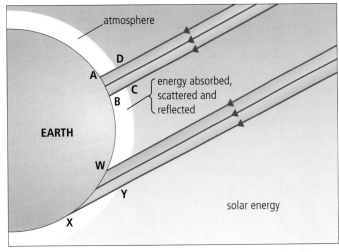

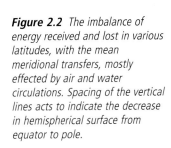

Figure 2.1 Variation in solar energy received per unit area of surface at different latitudes is shown by comparing **AB** and **WX** and atmospheric distances at **YX** and at **DA**.

The sun's elevation varies with latitude, seasons and throughout the day. Rays falling obliquely to the surface give less energy per unit area than those striking vertically. The energy is spread over a larger area (Figure 2.1). Passage through the lower atmosphere also varies with latitude – a longer passage making for more absorption and scattering.

The energy received also varies with seasonal changes in hours of daylight. At the Equator the longest period of continuous daylight is 12 hours and the elevation of the noonday sun is high throughout the year. Within the Arctic and Antarctic Circles there is continuous mid-summer daylight, but the noonday sun is never within 40° of the zenith. Nevertheless, during mid-summer continuous insolation produces a greater daily input of energy than is received on the Equator at that time. By contrast, during the long winter period of darkness energy continues to be lost from the surface, so that, all else being equal, the temperature falls progressively.

Elsewhere over the globe are various combinations of shorter winter days with the sun relatively low in the sky and longer summer ones with a higher noonday sun. Figure 2.2 shows the combined effects of length of day and the elevation of the noonday sun on the energy received at the outer limits of the atmosphere in different latitudes.

Energy redistributed – but affected by local conditions

Figure 2.2 stresses that higher latitudes would become progressively colder without energy transferred from the tropics and sub-tropics. In reality, winds and ocean currents continuously redistribute energy. However, the unequal

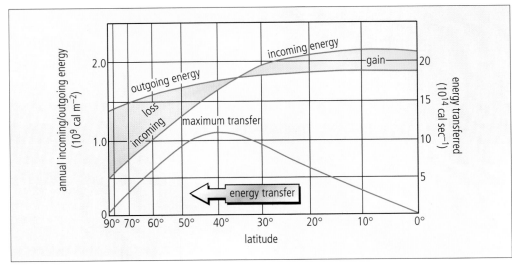

Figure 2.2 The imbalance of energy received and lost in various latitudes, with the mean meridional transfers, mostly effected by air and water circulations. Spacing of the vertical lines acts to indicate the decrease in hemispherical surface from equator to pole.

Figure 2.3 *Mt McKinley at 63°N in the Alaskan Range, where ice from the snow-covered mountain glaciers builds up the apron of a piedmont glacier.*

distribution of surface energy is not a gradual variation from low latitudes to Polar regions. The location of the hottest surfaces varies seasonally, while land and water absorb and radiate energy at different rates, as do innumerable surface materials and the varied topography.

Chapter 3 considers Earth's climatic regions, receiving seasonal solar energy according to their latitude. But the climatic conditions experienced by their peoples depend on a combination of other influences. For instance, the local air humidity and temperature are affected by vapour received from the surface, from lakes, rivers, floods and vegetation. They will also respond to local relief – to variations in aspect and altitude.

The landscapes in Figures 2.3 and 2.4 contrast the effects of energy supplied to Alaska over a few months by a low-angled sun with those in Sri Lanka, where during the year the noonday sun is never far from the vertical; but in fact they indicate more than that.

With humid air from the ocean there is a heavy snow cover about Mt McKinley, though over the years the amount of snow, and thus the extent of the piedmont glacier, varies. It tends to be more extensive during warmer periods and less during colder ones (p. 82). While in this coastal region of Sri Lanka the humid lower air receives a continuous supply of water vapour from rivers, flooded padi fields and the ocean. Despite higher temperatures this makes for condensation, cloud formation with the release of latent heat energy and, thus, instability and rainfall.

In each case the reception of a specific amount of solar energy, according to latitude, is only one factor affecting the local weather conditions. When considering the mean conditions of temperature, pressure and prevailing winds shown by the following maps, it is essential to realise that for many reasons, including those expressed above, such averages conceal innumerable local variations in weather and climate.

Figure 2.4 *The west coast of Sri Lanka where teak trees give way to numerous coconut palms fringing the river, creeks and flooded padi fields.*

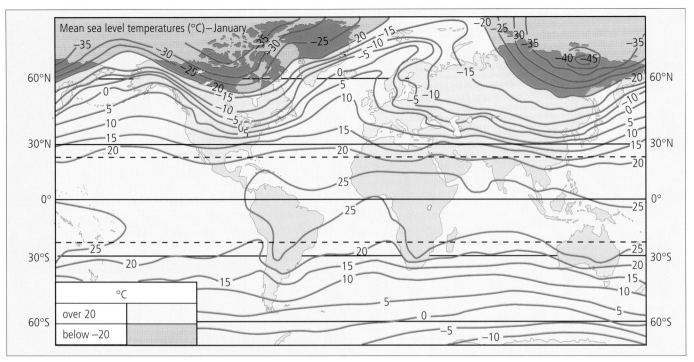

Figure 2.5 *The contrast in mean temperatures between the two hemispheres is striking. In the cold northern 'land hemisphere' maritime influences extend northward, so that at each latitude there are large differences in mean temperature between air over oceans and air over continental interiors.*

World-wide contrasts in seasonal temperature

In Figure 2.5 the **January isotherms** indicate overall energy losses from the northern landmasses and the poleward transport of energy by warm waters in the northern oceans. While in the southern hemisphere they emphasise both the latitudinal influence and a uniformity imposed by the oceans.

The **July isotherms** (Figure 2.6) point to the contrast between summer heat in the interior of the northern landmasses and energy loss during winter. They also show that in the northern hemisphere arid tropical and sub-tropical regions affected by air subsidence are very hot indeed. Yet overall the contrasts between mean air temperatures over

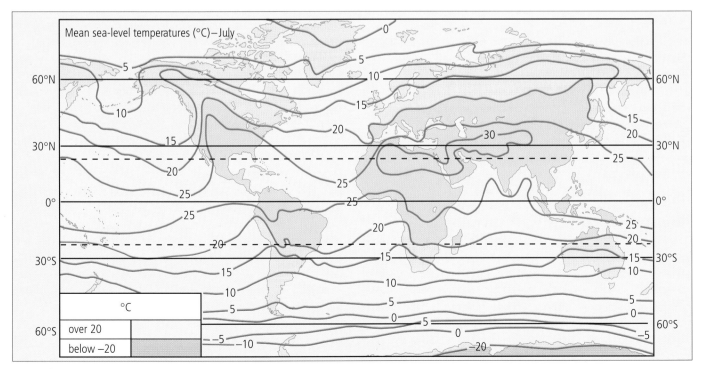

Figure 2.6 *There is an overall decrease in mean temperature poleward. In the northern hemisphere the mean temperatures over continental interiors contrast with those in January. Notice the influence of cold ocean currents to the west of the land areas.*

oceans and land areas are less than in January.

Together the maps stress the large **mean annual temperature range** of the northern continental interiors compared with the small annual range throughout the southern hemisphere, where the east–west trend of isotherms reflects the decrease in insolation received from equatorial to polar latitudes. Again, these mean figures mask innumerable local differences. The apparent uniformity in mean temperatures within the tropics conceals immense variations, within the Andean mountain system for example (p. 57), and hides the local effects of energy absorbed by the surface and energy reflected, especially by ice and snow.

Inset 2.1
A caution

It is essential to understand what isotherm maps tell us and what they conceal. **Mean sea-level isotherms**, based on calculations of what the temperature would be at sea-level, eliminate many relief effects, but highlight such influences as prevailing winds, ocean currents and 'continentally'. However, 'mean' figures can mislead.

A mean temperature of 27°C for a particular day, calculated by adding maximum and minimum readings and dividing by two, may conceal a maximum of 40°C and a minimum of 14°C, or a maximum of 32°C and a minimum of 22°C.

A **mean monthly temperature** of 20°C for a particular July, the total of daily means divided by 31, may hide a spell when daily temperatures soared to 30°C, or a cool one when they failed to reach 15°C. As the mean July temperature for that location may be an average of 30 consecutive July means, the limitations of such a figure are obvious. Remember, too, that **climate as well as weather varies over the decades.**

Air and ocean currents transfer energy

About 60% of the energy transferred from lower to higher latitudes is through air circulation in the lower and upper parts of the troposphere. Apart from air's kinetic energy of movement, it acquires water vapour from ocean surfaces, which releases heat energy on condensation. Ocean currents, whose movements are mainly caused by wind drag, transfer about 25%. Variations in sea-water density also play a role by creating slow circulations from surface water to ocean deeps between polar and tropical regions.

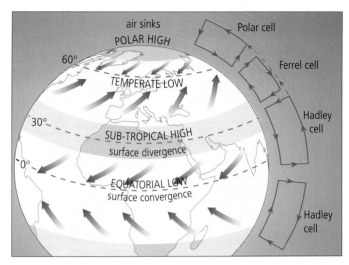

Figure 2.7 *Air pressures and surface winds related to three simplified cells of circulation.*

Global air circulation – simple patterns

Global air circulation is not simply a matter of large temperature differences causing air to move in two convection cells from the Equator to Poles. Apart from the Coriolis force deflection, a mass of air acquires various properties as it moves, or remains static, over a surface. As it is heated, cooled or moistened, there are changes in density and air pressure, which affect its horizontal and vertical movements. In higher latitudes especially, surface air pressure and winds are also affected by exchanges with upper-air currents – high altitude jet streams (p. 28). Even so, despite these ever-changing atmospheric conditions, certain air movements occur regularly enough for us to recognise the global patterns of pressure and winds shown in Figures 2.8 and 2.9.

As long ago as 1735, George Hadley theorised about a global convection system set in motion by a tropical heat source. Then in 1856, William Ferrel put forward a circulation pattern for each hemisphere, based on three cells (Figure 2.7). It was an over-simplification, but a starting-point from which to examine more complex vertical and horizontal air movements.

In each hemisphere he recognised the **Hadley cell**, where air rises in equatorial regions and moves poleward at high altitudes, deflected by the Coriolis force as a westerly flow. About latitude 30° the air subsides, some returning to low latitudes as the easterly flow of the Trade Winds, other air moving poleward as surface westerlies. He pointed to a **Polar cell** where cold, dense air returns towards lower latitudes as easterlies. Between the Hadley and Polar cells their outflowing surface winds clash and circulate in his **Ferrel cell**.

This early model was used to indicate where rising air makes for considerable precipitation, and where sinking air creates high pressure zones with arid conditions, notably the hot deserts.

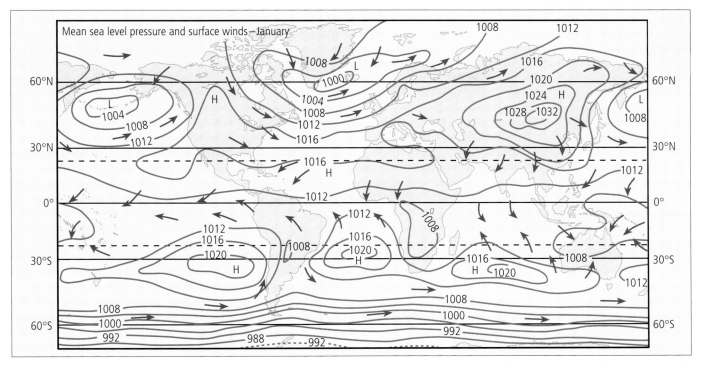

Figure 2.8 The January isobars emphasise the thermal high over central Asia, with outflows during the Asiatic dry monsoon. Notice the convergence of the tropical easterlies and the position of the temperate westerlies.

Mean sea-level pressure and seasonal winds

The average pressure distribution and resulting winds are broadly those of the early model: tropical easterlies converge; sub-tropical high pressure cells over the oceans extend over land areas, especially during summer; westerlies are strong in the middle latitudes, especially in the southern hemisphere, and are deflected across pressure gradients. However, in the northern hemisphere in particular, these 'patterns' are disturbed by landmasses, where the seasonal pressure variations influence Asia's monsoonal inflows in summer and outflows during winter. Pressure remains high over Antarctica. It is more variable in the Arctic, where highs are less persistent, though outflows of cold air influence weather in the higher middle latitudes.

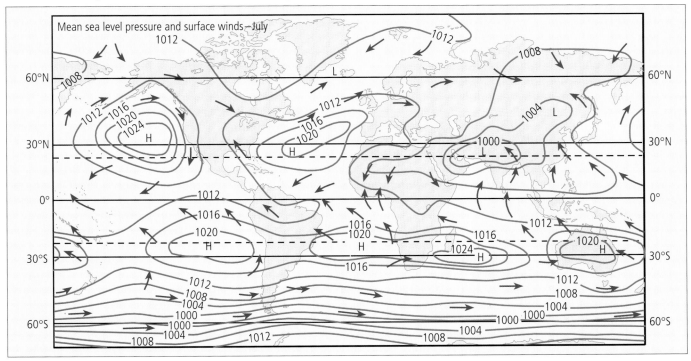

Figure 2.9 The July isobars indicate the low pressure over central Asia, with inward flows during the Asian wet monsoon, and shows the persistent westerlies encircling Antarctica.

QUESTIONS

1 Consider the basic differences of land and water surfaces in relation to their ability to absorb and transmit energy. Point out how the mean seasonal temperatures shown in Figures 2.5 and 2.6 reflect such differences.

2 In the Hadley-Ferrel convection system the cells show vertical and latitudinal airflows. Why do prevailing surface winds tend not to follow lines of longitude, but blow as 'tropical easterlies' or 'mid-latitude westerlies'?

3 How do the sub-tropical zones of high pressure influence these prevailing winds?

UPPER AIR AFFECTING SURFACE WEATHER

Variability of surface weather

There is now a more realistic view of global air circulation and causes of particular weather patterns. The maps of average conditions mask very variable weather in many parts of the world; notably in the middle latitudes where eastward-moving depressions are accompanied by day-to-day variations in wind direction. The apparently persistent zones and cells of high pressure in the southern hemisphere are not stationary systems, but a westward-moving procession of highs and troughs of low pressure (p. 34).

Air accumulates in certain source regions and remains long enough in contact with the surface to transmit local properties of energy and humidity into its mass. Such air masses move from their source in response to pressure gradients and affect weather in the regions they invade (p. 36).

Maps of average pressure and corresponding winds tell little of changing weather, of its causes or of air masses involved. It is necessary, therefore, to examine the actual three-dimensional air movements and the role of upper-air flows on weather conditions and climatic variations.

Upper-air flows and the Hadley cell

The Hadley cell is a means of transferring energy from low latitudes, though the mechanism is more complex than Hadley and Ferrel envisaged. In low latitudes energy is supplied to the upper air by convection and the release of latent heat. As the air moves poleward, the lack of friction and the increasing Coriolis force cause it to turns eastward as a geostrophic wind of increasing speed. Between 23° and 30° latitude it becomes part of strong upper westerlies. Above some regions the core of these westerly air currents is particularly concentrated, as an energetic, meandering, **sub-tropical jet stream**, some hundreds of kilometres wide (Figure 2.11).

High-level air continues to feed into these streams, though the speed of the streams is limited by turbulence and energy losses. This makes the air subside, which then feeds a high pressure zone at the surface. Also as air in loops of the jet stream moves equatorward, it slows and leads to a piling up – convergence – which causes some of the upper air to sink (p. 30).

Most subsidence takes place over the westward parts of the sub-tropical high pressure zones, where such sinking air effectively blankets that rising by convection and acts to maintain the aridity of the hot desert surfaces.

Figure 2.11 shows mean seasonal latitudinal positions of these upper westerlies (W), whose speed varies about the globe. Their paths and the extent of looping of the jet streams are actually very variable. Over Asia there is a less continuous flow during the northern summer. In fact where pressure is high above the heated Tibetan plateau (p. 38) a southward outflow develops, deflected to form strong upper easterlies (E).

The air sinking in the sub-tropics diverges at the surface. Easterly Trade Winds return to low latitudes, completing the Hadley cell; but the surface air moving poleward becomes part of circulating pressure systems, rather than a simple westerly flow.

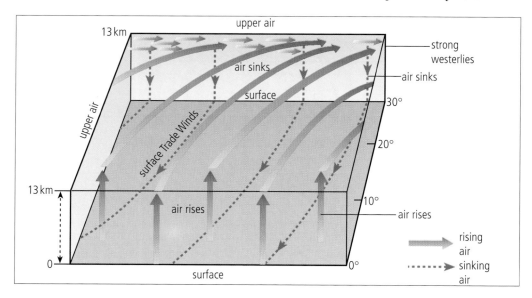

Figure 2.10 The Hadley cell in the northern hemisphere, with upper westerlies in the sub-tropics and the surface return of north-easterly 'Trade Winds'.

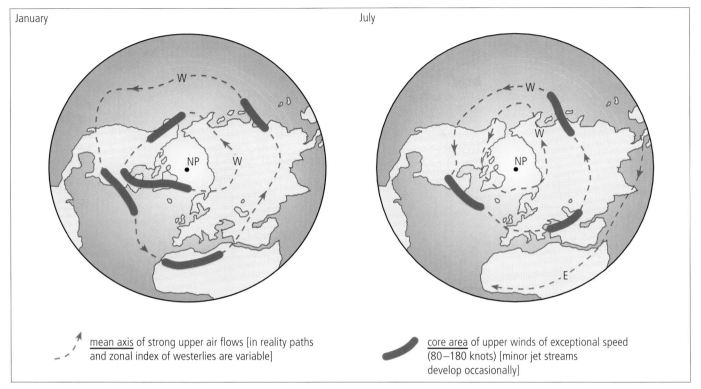

January   July

mean axis of strong upper air flows [in reality paths and zonal index of westerlies are variable]

core area of upper winds of exceptional speed (80–180 knots) [minor jet streams develop occasionally]

Figure 2.11 A generalised view of the strong upper westerlies, with core areas of powerful jet streams, and showing the upper easterlies during the northern summer.

Upper westerlies of the higher latitudes

In these latitudes surface low pressure systems move eastward over the oceans and western parts of the continents. They, and intervening areas of high pressure, are closely related to upper-air temperatures, pressures and air movements.

North-south temperature contrasts are much greater than in the tropics. As the consequent differences in pressure cause the upper air to flow poleward, it is strongly deflected eastward by the Coriolis force into fast-moving, looping **upper westerlies** with **polar front jet streams** at their core (Figure 2.11). Their swing varies with the angular momen-tum of the air stream. High physical barriers, such as the Rocky Mountains, decrease the speed of rotation of the upper westerlies. But then their wave amplitude increases as they continue eastward in long loops – Rossby waves.

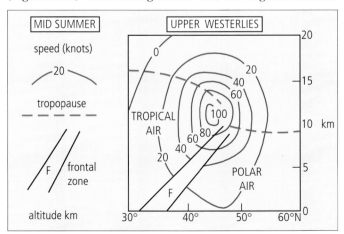

Figure 2.12 A core of upper westerlies during the northern summer, with a zone where fronts develop between polar and tropical air. Notice variations in altitude of the tropopause.

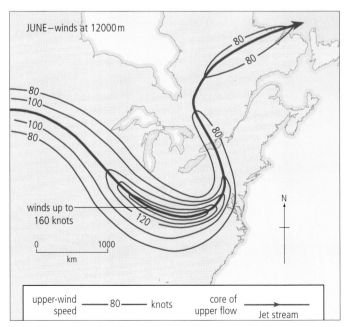

Figure 2.13 The looping upper westerlies in the lee of the Rocky Mountains, with a particularly strong jet stream about latitude 40°.

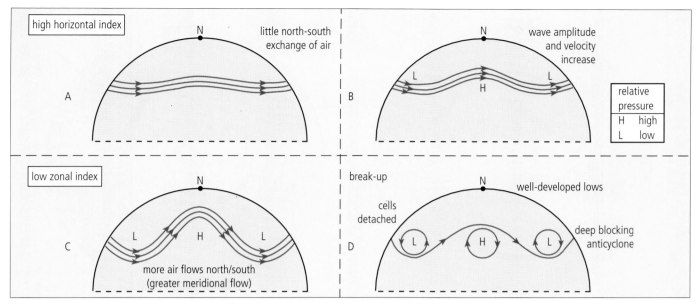

Figure 2.14 In **A** and **B** there is a strong zonal flow with gentle wave motion. In **C** the amplitude of the waves has increased, with distinct high and low pressure centres. In **D** a 'cut-off' has produced a stationary 'blocking' anticyclone that moves only slowly eastward, and the wave amplitude has decreased.

Inset 2.2
Wave formation

- In all fluids, including the upper air, the mechanics of stream flow tend to create meanders of varying amplitude.

- As a column of relatively light warm air rises from the surface, a ridge develops in the upper air; while above a sinking column of cold air there is an upper-air trough. Thus at a particular altitude various ridges and troughs may divert upper-air streams.

- Upper air moving poleward speeds up; as in the northern hemisphere, where in an upper wave it loops anti-cyclonically northward. As the air then curves cyclonically towards lower latitudes, it slows.

A strong west-east (zonal) flow with little meandering is more usual in winter – a flow with a **high zonal index** – for then there is strong contrast between low pressure above cold polar air and higher pressure in lower latitudes. Later there tends to be more pronounced looping in the upper westerlies, with troughs and ridges accentuated and centres of high and low pressure. Occasionally a stationary high pressure zone is cut off (Figure 2.14D), creating a blocking anticyclone in the weather system.

Upper-air currents affect surface weather

As the upper westerlies loop poleward about a ridge, the air speeds up and diverges. This allows air beneath to rise and makes for low pressure near the surface (Figure 2.15). As the air then turns equatorward and slows, it converges and piles up, which causes air to descend and create high pressure below. Surface weather is often closely related to the nature of upper-air flows. However, this is a two-way situation –

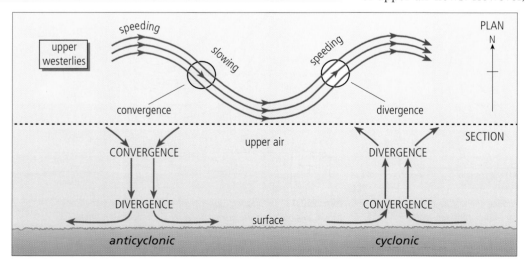

Figure 2.15 As air in upper waves swings northward with increasing speed, it diverges and so draws up air from below. Swinging southward its speed decreases and convergence causes air to sink.

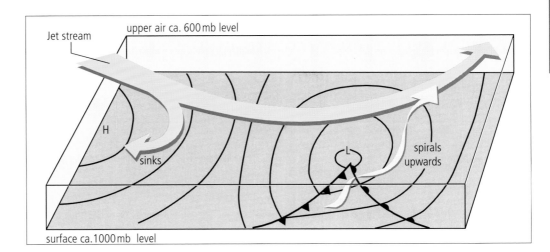

Figure 2.16 *Air convergence in the upper westerlies acts to strengthen an anticyclone, while divergence causes air to spiral upward, making for a deepening depression near the surface.*

energy injected into the upper air by strong convection or upsurges at fronts (p. 32) affects the flow of the upper westerlies. Periodic variations in solar energy inputs also affect both upper-air and surface flows, with appreciable influence on `climatic change.

Mid-latitude depressions (cyclonic systems)

As air subsides over polar regions there are occasional outward surges of cold air. These may meet warm air from the sub-tropics at a **polar front** (Figure 2.27). As this cold air surges into warm air at the surface, forming a **cold front**, it forces the lighter air to rise and pressure falls. A cyclonic system tends to form around the centre of low pressure – a **depression**. If divergent upper air allows the warm air to continue to rise, surface pressure falls further.

As surface air spirals inward to the developing centre of low pressure, the **cold front** advances, narrowing a sector of warm air. The whole system moves eastward, with the leading edge of the warm surface air advancing against colder air as a **warm front**.

Ahead of this front the warm air moves up and over a gentle retreating slope of cold air and may continue to rise towards the upper westerlies.

At the rear of the depression the cold dense air, fed perhaps by convergent air in the looping jet stream, tends to advance more rapidly than the system as a whole and so continues to narrow the warm sector. Finally it may lift the warm air clear of the ground. Such **occlusion** begins near the low pressure centre and develops outward. Figure 2.17(2) represents a fully developed depression. Notice how the isobar spacing shows different pressures of cold and warm air and the sharp angle across a front.

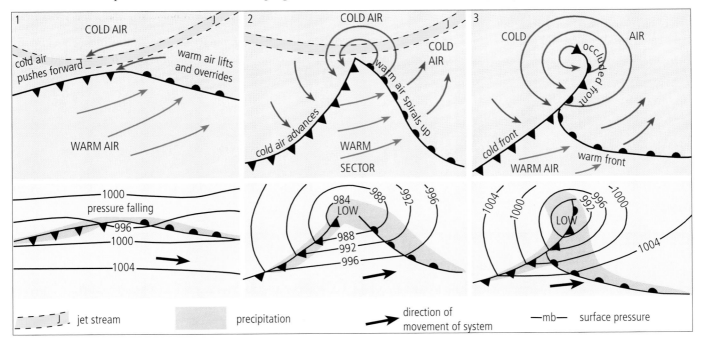

Figure 2.17 *Upper-air flow related to a deepening depression, with cold air advancing at the surface, undercutting warm air, narrowing the warm sector, and lifting the warm air clear of the surface (occlusion).*

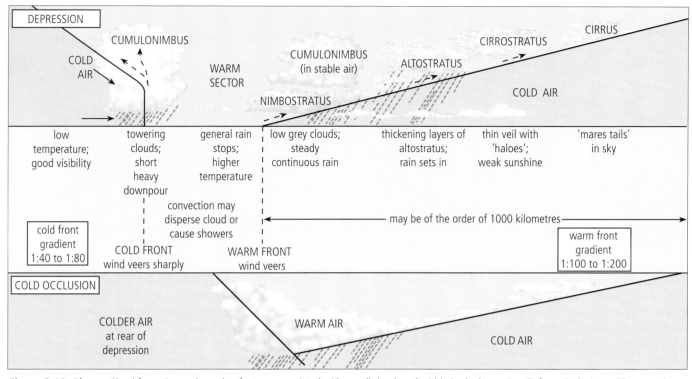

Figure 2.18 diagram labels:

DEPRESSION

CUMULONIMBUS

COLD AIR

WARM SECTOR

CUMULONIMBUS (in stable air)

NIMBOSTRATUS

ALTOSTRATUS

CIRROSTRATUS

CIRRUS

COLD AIR

| low temperature; good visibility | towering clouds; short heavy downpour | general rain stops; higher temperature | low grey clouds; steady continuous rain | thickening layers of altostratus; rain sets in | thin veil with 'haloes'; weak sunshine | 'mares tails' in sky |

convection may disperse cloud or cause showers

may be of the order of 1000 kilometres

cold front gradient 1:40 to 1:80

COLD FRONT wind veers sharply

WARM FRONT wind veers

warm front gradient 1:100 to 1:200

COLD OCCLUSION

COLDER AIR at rear of depression

WARM AIR

COLD AIR

Figure 2.18 Above: *Cloud formation and weather features associated with a well-developed mid-latitude depression. **Below**: Occlusion, with warm air overlying the denser colder air that has advanced at the rear of the depression.*

Sometimes secondary waves develop on the cold front, so that secondary depressions slowly encircle the primary one. In the British Isles these give long spells of changeable weather, with high pressure ridges between giving a dry day or two.

The first signs of an approaching depression are usually high, feathery 'mare's tail' cirrus, perhaps several hundred kilometres ahead of the warm front, or a sun halo in a thin veil of cirrostratus. Then thicker altostratus, with occasional light rain, heralds a period of persistent rain from the dark nimbostratus ahead of the front itself.

As the warm front passes, the wind veers and the temperature rises. Humidity remains high and there may be scattered convection showers. At the cold front, where warm air is vigorously forced up at a steep angle, huge cumulonimbus clouds may produce a brief torrential downpour before the sky clears and the temperature falls, with cold air moving in from a poleward direction.

Thus the 'Ferrel cell' is a zone with eastward-moving depressions along polar fronts, sometimes checked by 'blocking highs', and interrupted during the northern hemisphere winter by outsurges from cold air masses over the continental interior.

Figure 2.19 *Brighter colder air follows the towering cumulus and torrential rain at the cold front of a depression that has passed over Vancouver from the west.*

Synoptic weather maps

In temperate latitudes the sequence of eastward-moving depressions, blocked at times by well established highs, makes for changeable weather in the British Isles. Weather maps provide much information about current weather features and allow forecasts of likely changes, as in Figures 2.20 and 2.21.

Figure 2.20 is a simplified map for a day in early autumn, at a time when the upper westerlies had a high zonal index.

QUESTIONS

1 Look carefully at the isobars, temperatures (°C), wind directions and position of the fronts shown in Figure 2.20. Describe and account for weather conditions at A, B and C and the differences between them.

2 Account for the differences in wind directions over the North Sea and the eastern Atlantic.

3 What is meant by 'zonal' in relation to the upper westerlies?

The simplified maps in Figure 2.21 show pressure conditions that helped to give the British Isles a period of hot summer weather.

QUESTIONS

1 Figure 2.21(A) shows air pressure during Britain's hot sunny period in June when the sky was mostly clear and that in (B) when the late August weather was very hot, humid and less comfortable. Two 'heatwaves', but with different weather – why?

2 During June some places were relatively cooler, and some had sunny intervals and mainly light cumulus clouds. Suggest where these were and why.

3 There are certain similarities in wind direction in both A and B. Comment on this with reference to the pressure distributions.

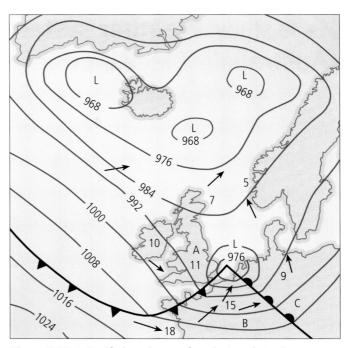

Figure 2.20 *A simplified weather map for a day in early October – temperatures in °C*

In all these maps there is low pressure about Iceland, a principal feature of the mean North Atlantic pressure pattern. In contrast to the British Isles it is less influenced by the continental air mass sources shown in Figure 2.28.

In the southern hemisphere there is also a procession of pressure systems across the wide oceans, bringing the sequence of changing weather to southern Australia illustrated in the following case study.

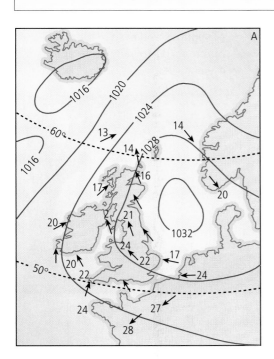
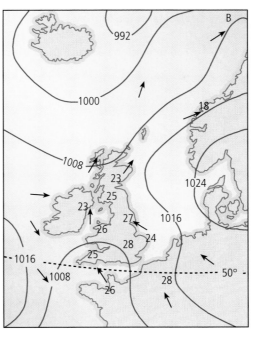

Figure 2.21 *Meteorological conditions during two short 'heatwaves' in the summer of 1999: **A** during June; **B** during late August.*

Case study: **Weather systems on the move – Australia's 'Black Wednesday'**

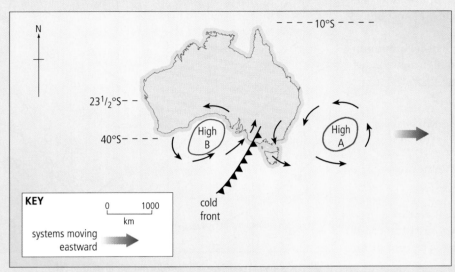

Figure 2.22 *Anticyclones travelling eastward to the south of Australia, with abrupt changes in wind direction in the trough between.*

The pattern of winds prevailing during a particular season gives a very generalised picture of climatic conditions, with little indication of what are often abrupt changes in the weather from place to place. This is the case with the broad flow of westerlies across the wide oceans of the southern hemisphere, for within them successions of anticyclones also move eastward, covering some 500 km per day, and between these are troughs of lower pressure.

Figure 2.22 shows that to the south of Australia air is moving anticyclonically around these high pressure systems. In the case of A this causes winds to move from the interior over the Murray Basin and south-eastern coastlands; while in response to the approaching anticyclone B cooler air from the southern ocean is drawn northward towards the Murray.

In summer in particular there is a great difference in temperature between air from the heated interior and that from over the cool southern waters, so that within the intervening trough there is a marked cold front. This is seen in Figure 2.25, which shows the eastward-moving trough on 8 January 1969 with associated winds, whose devastating effects gave the day the name 'Black Wednesday'.

Early that Wednesday strong northerlies with a temperature of 45°C set in ahead of the approaching front, stirring up dry surface air. The sun became masked by red-brown dust and visibility closed in. Aware of the hazard, authorities declared a total fire ban. Nevertheless a single fire can spread alarmingly, and soon sparks carried by the strong wind ignited vapour from the eucalypts, whose whole stands burst into flames – like a series of torches across the landscape. The flames spread through woods and crops, destroying livestock, farmhouses and vehicles. Local fire-engines strove to limit the damage. As the air moved southward, there were fires about Melbourne, and even in the south-west outbreaks destroyed houses and claimed lives.

Figure 2.23 *In the early morning a mass of dust carried by the hot dry northerlies approaches Seymour, in the south of the Murray Basin.*

Figure 2.24 *Dust-laden air from the north moves across wheatfields inland of Seymour, across a landscape about to be devastated by fire.*

Figure 2.25 *The location of the eastward-moving trough in the early hours of Black Wednesday. There was striking difference in temperature and humidity between air to the east of the front and air to the west.*

This was Black Wednesday – yet remarkably, late in the day, as the front passed, the very strong 'southerly buster' set in, and within an hour the temperature had dropped from 45°C to 15°C!

Since then other dust-storms have blanketed these south-east coastlands, for these travelling pressure systems are a major synoptic feature of the so-called 'west wind belt'. This once again demonstrates the need to appreciate the effects of changes in pressure conditions and air stability – mean climatic statistics of an area under consideration mask so much and can be misleading.

Figure 2.26 *Late in the day, as skies cleared, a local fire-engine continued to damp-down the still smouldering fields near Seymour. Maintaining fire services is a high priority for small communities in these extensive areas of wheat, sheep and cattle farming.*

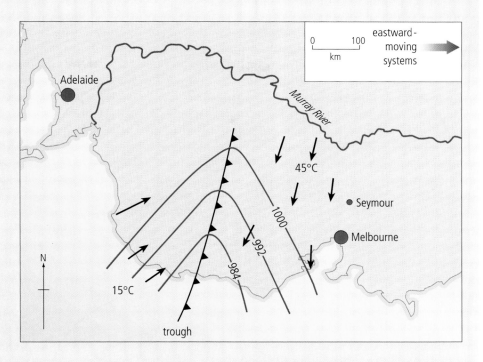

QUESTIONS

1 The 'southerly buster' is seen here as relieving devastation over a huge area. But suggest why at times it brings towering cloud and violent thunderstorms to these southern regions.

2 Why are such regularly travelling anticyclonic cells not characteristic of temperate latitudes in the northern hemisphere?

3 Hearing that high pressure is becoming established over the British Isles, in summer one expects clear skies and hot sunny days, in winter one fears frosts and fog. Why exactly are such conditions likely to develop?

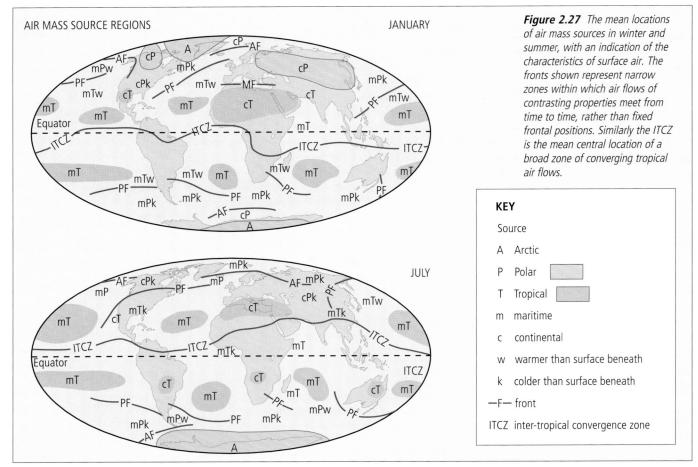

AIR MASS SOURCE REGIONS   JANUARY

JULY

Figure 2.27 *The mean locations of air mass sources in winter and summer, with an indication of the characteristics of surface air. The fronts shown represent narrow zones within which air flows of contrasting properties meet from time to time, rather than fixed frontal positions. Similarly the ITCZ is the mean central location of a broad zone of converging tropical air flows.*

KEY

Source

A	Arctic
P	Polar
T	Tropical
m	maritime
c	continental
w	warmer than surface beneath
k	colder than surface beneath
—F—	front
ITCZ	inter-tropical convergence zone

Air masses – sources and outflows

In certain parts of the world air accumulates and remains stationary long enough to acquire specific characteristics from the surface, which are distributed slowly through its mass. Figure 2.27 shows the mean seasonal location of air mass sources, with fronts representing narrow zones where air from such sources meet from time to time.

The ITCZ (inter-tropical convergence zone), moving with the 'overhead sun', indicates the mean location of converging flows of air from cT and mT sources – a broad fluctuating convergence zone, rather than 'frontal', for the opposing flows do not have the sharp contrasts in temperature and density as those confronting each other in higher latitudes. Nevertheless within the zone there is turbulence with thundery storms as inflowing humid mT air passes over strongly heated surfaces. Thus the summer advance of the ITCZ, heralding the arrival of unstable mT air, is welcomed by such semi-arid sub-tropical lands as the Sahel, south of the Sahara, whose climate is apt to be dominated by dry sinking air from the north. However, the amount of rain from such mT air varies considerably from year to year (p. 87).

As the dry Saharan air moves outward over the Atlantic, it gains moisture but passes first over a cool surface, and so as mTw air it is warmer than the surface and remains stable for a long time. Where mT air is cooler than the surface the symbol k is used. But even mTk air may remain stable over tropical oceans, which lack the strong surface/air contrasts of the temperate regions – witness the

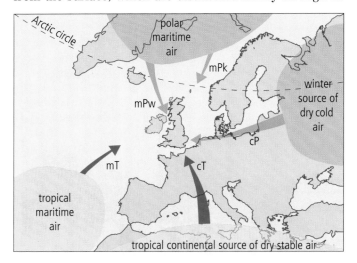

Figure 2.28 *Air mass sources that affect the British Isles. Air properties are likely to be modified by passage over the ocean, for often the air does not arrive direct, but after circulation about a particular pressure system. Flows of mT air are the most frequent during the year, those of cP and cT air much less so.*

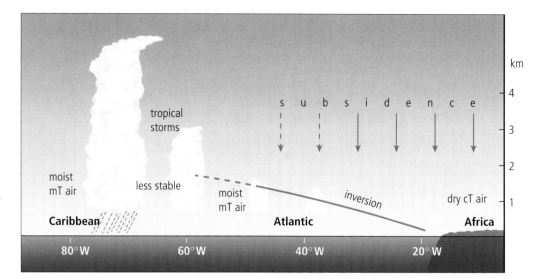

Figure 2.29 *Subsiding air in the sub-tropics remains stable over the eastern Atlantic. The Trade Winds pick up moisture, so that further west potentially unstable air produces lines of towering cumulus, with occasional rainstorms.*

sunny weather with scattered cumulus of the Trade Wind zones (Figure 2.29). Nevertheless mT air is potentially unstable, as shown by the effects of summer inflows in the Asian wet monsoon systems, and as it passes over a tropical ocean there may be stronger uplift in response to upper-air divergence, probably triggering a tropical cyclone (p. 39).

Figure 2.27 shows that dry, stable cT air from the Saharan sub-tropical high pressure affects southern Europe during summer, when its advance over the Mediterranean makes for lengthy drought. In winter, however, lows may form at a 'Mediterranean front', where surges of cold cP air meet inflowing mT air. Active depressions also occasionally move in from the Atlantic.

Figure 2.27 also shows the winter source of very cold cP air in the interior of Euro-Asia, where subsiding air remains as a cold mass, with low humidity and temperature inversion in the lower atmosphere. Where this cP air passes outward over warmer seas it acquires heat and moisture, and becomes less stable maritime mP air. The Arctic and Antarctic regions are other sources of very cold cA air.

MONSOON SYSTEMS – TROPICAL CYCLONES

The monsoons in southern Asia

The term 'monsoon' derives from a word for 'season'. Thus, as seasonal pressure changes cause a reversal from dry winter conditions to the summer inflows of moist air, we refer to a 'dry monsoon' and a 'wet monsoon'.

During winter there are strong upper westerlies south of the Himalayas and air subsides over the northern plains of the Indian sub-continent. During this **dry monsoon**, from November to February, pressure remains high, though occasional weak depressions move eastward across the far north-west. Days over the plains are very warm, nights crisply cool. Surface air flows as dry north-easterlies over the Indian peninsula towards the ITCZ, now south of the Equator.

From mid-March, as the temperature rises rapidly, pressure falls over the northern plains. But a declining high still separates the dry air of this developing low from the far south, where warm humid mT air from over the southern ocean is creating scattered storms.

Figure 2.30 *One of a series of tropical rainstorms in unstable air from the Atlantic passing into the Caribbean, north of Antigua. Notice the anvil effect above the column of cumulus cloud. The lower clouds show a regular level of condensation and form individual billowing masses.*

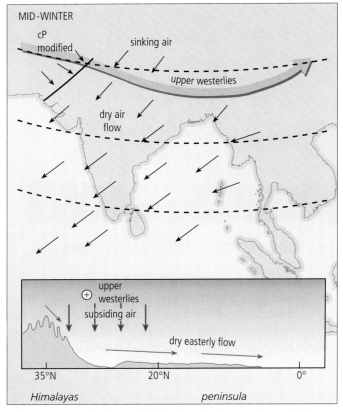

Figure 2.31 *During winter subsiding air maintains a south-easterly flow over the peninsula. Weak lows from the west occasionally bring light showers to the north-western plains.*

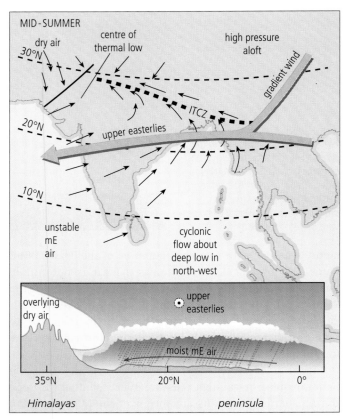

Figure 2.32 *The moist inflow brings heavy rain, especially to the north of the Bay of Bengal. Drier conditions prevail over the far north-west. High pressure above the Tibetan plateau creates a southward flow, deflected to form the strong upper easterlies.*

Suddenly at the end of May the mT air surges into the now deep thermal low and slowly advances as far as 30°N, bringing heavy rain. As much of this moist low-level inflow has crossed the Equator, this is sometimes termed mE air. The 'burst' of this **wet monsoon** occurs so regularly that agricultural practices are geared to its arrival, though occasionally a long delay causes crop failures. However, during the summer months the occurrence and intensity of rain varies, partly due to shallow disturbances in the inflowing mT air. In some years there are unusually long dry spells amid the rains.

Topography greatly affects the amount of precipitation received. Mean annual rainfall decreases from the heavy monsoon rainfall to the north of the Bay of Bengal, westward across the plains, to where in the north-west subsiding air overlies the shallow summer inflow and blankets vertical movements of moist air – thus leaving the hottest parts of the sub-continent the driest. The rainy period lasts until mid-September in most places, when subsiding air gradually re-establishes itself in the north. As the ITCZ retreats southward, occasional tropical storms still affect parts of the south.

During spring the upper westerlies move north of the Himalayas, and from early summer an easterly jet stream develops to the south (Figure 2.32). The Tibetan plateau, over 4000 m above sea-level, is now a high altitude heat source, establishing high pressure in the overlying upper air. With a north–south gradient this upper air flows southward, deflected to the west as an **easterly jet stream** some 15 km above south-east Asia, and continuing over the Middle East to the southern Sahara. Over India, its meanders cause air to rise on the northern side and descend to the south. This, and its variation in strength and breadth, provide another reason for changeable weather during the wet monsoon.

The monsoons of eastern Asia

During summer humid air moves in mainly from the east and south-east, a weak wet monsoon compared with that of south-east Asia, which itself affects part of the south-west. However, the coastal south-east, especially the far south with its prominent relief, receives about a third of its summer rain from typhoons that develop in the humid inflow from the Pacific.

To the north of this huge area such monsoonal inflows bring less rain; very little indeed to the far interior. However, north-eastern coastal regions also receive rain from fronts where warm mT air meets cool mP air from over northern waters.

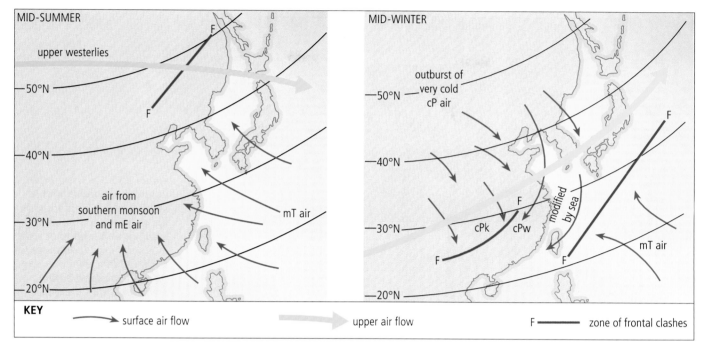

Figure 2.33 *Climatic effects of East Asian monsoons vary with latitude. North of latitude 40° the moist inflow during summer only occasionally penetrates far inland. In winter the cold continental outflow is modified by ocean influences as it circulates southward.*

During winter cold, dry, stable air arrives in outbursts from central Asia, carrying much wind-borne dust (p. 50). The north-east is bitterly cold with periods of snow in near-coastal regions. Some of this air swings clockwise around the continental high, gaining moisture from the Pacific, and as it clashes with mT air, it brings cloud and precipitation to the eastern coastal lowlands of central and southern China. Further inland as air from the high Tibetan plateau moves eastward, it warms on descent, so that the mid-western interior, which is also sheltered from the cold northerlies, has much mild sunny weather.

Tropical cyclones (typhoons)

Summer inflows bring rain to other continental regions besides south-east Asia and the Indian Ocean. Intense cyclonic storms affect Central America, south-eastern USA and northern Australia. These tropical cyclones are systems that, gaining heat and moisture from surface waters above 27°C, develop almost symmetrical swirls of hurricane-force winds around a calm centre. They form over oceans in the mid-tropics, though seldom within seven degrees of the Equator where the Coriolis force is slight, and tend to affect eastern parts of the oceans during late summer and early autumn (Figure 2.34).

Low pressure troughs, with clusters of tall cumulus clouds indicating strong vertical development, frequently cross the Pacific and Atlantic. In some cases upper-air divergence increases the uplift, so that the rapidly rising air rotates to form a tight cyclonic vortex. Energy released by condensation boosts the rise of towering cumulonimbus, with very

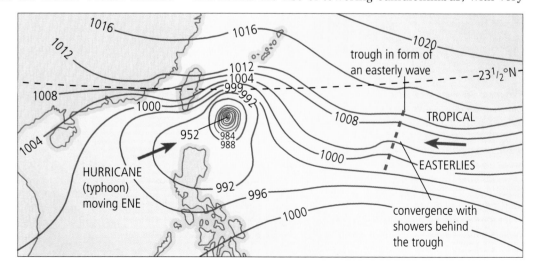

Figure 2.34 *A typhoon moving eastward towards Taiwan, with a following wave disturbance in the Trade Wind flow over the western Pacific.*

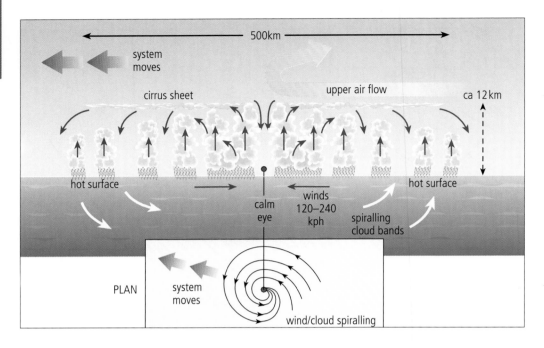

Figure 2.35 *The system moves westward, with individual cloud bands spiralling inward about the eye of the cyclone. As it deepens, ascending air is removed by upper-air currents some 12 km above the surface. As long as it moves over the very warm ocean, surface energy released by condensation keeps it active.*

heavy rain about the storm centre. There amid the spiralling cloud masses air descends and warms, giving a clear 'eye'. As the ascending cloud reaches the tropopause, it spreads as a wide layer about the system.

Pressure gradients about the centre generate winds of great force, which can build up destructive tidal waves. As the centre itself passes over a location, there is a change in wind direction of 180°. The whole westward-moving system is often many hundreds of kilometres across, but accompanying rainfall, though torrential, varies considerably within the system, depending on location and the speed of passage.

Intense tropical cyclones develop at times in the Indian Ocean, causing tidal waves to devastate coastal lowlands, as in Bangladesh (p. 84), and in Orissa in 1999. Torrential rain may also inundate populous floodplains, as in Mozambique in 2000.

Tornadoes, vigorous up-spirals of air above a hot surface, are much smaller systems, some only 100 m or so across, but can also be extremely destructive, writhing their way across the countryside at some 50 kph, detaching and sucking up debris. In many cloud funnels up, then spreads as dense cumulonimbus. Where they move over the sea, water may be sucked up towards the overhanging cloud, as a **water spout**, then flung out again by centrifugal force, frequently with fish as well.

Figure 2.36 *Intense cyclonic storms occur mostly over the western parts of the oceans, towards the outer tropics. Nearer the Equator the Coriolis force is weak, so few storms such as these develop in the low latitudes. However, cyclones may create tidal waves that travel hundreds of kilometres to affect distant low-lying equatorial atolls.*

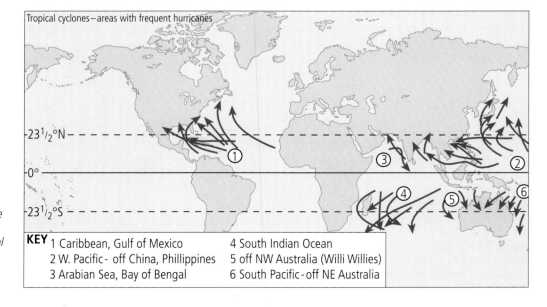

Case study:
Cyclones – transfers of energy

Figure 2.37 Rapidly rising cumulus indicates unstable air over San Juan, Puerto Rico.

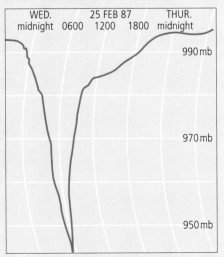

Figure 2.38 The passage of cyclone Elsie in February 1987, deepening as it picked up moisture from the Indian Ocean and crossed the West Australian coastline.

Caribbean turbulence

Cyclones emphasise that weather conditions involve exchanges of energy. Figure 2.37 shows the result of the transfer of water from the heated surface of the tropical Atlantic into water vapour in very warm air. Here, as the air moves over the land and rises by convection over the hot surface, condensation of the vapour releases more energy, which stimulates the rapid rise of this towering cumulus.

In this case the instability heralded the arrival two days later of an even more energetic system – a typical late summer Caribbean cyclone, whose eye skirted the island, yet transferred sufficient energy to the sea to wreck sizeable craft within this normally sheltered harbour.

Cyclone Elsie

Northern and north-western Australia experience intense cyclonic storms. In 1987 cyclone Elsie gathered energy from the sea surfaces to the north-east and for several days tore westward along the path shown in Figure 2.38. Diverted by an upper high, it deepened as it moved southward over the heated surface of the Indian Ocean. Thus acquiring additional potential energy, it then swung inland over Western Australia. With gusts of over 200 kph, it tore through the property of an isolated homestead, whose barograph recorded the trace shown in Figure 2.39. Throughout its course it missed the larger settlements of this sparsely populated part of the country, though hundreds of sheep and cattle were lost.

QUESTIONS

1 The path of cyclonic systems, typhoons or tornadoes, responds to wider pressure distribution.
   a Explain how this is the case in eastern Asia.
   b How does information from the maps in Figures 2.43 to 2.45 suggest their likely paths into and over eastern USA?

2 a Why are such rapidly revolving systems less likely to develop over oceans close to the Equator?
   b What upper-air conditions may favour the development of powerful cyclones?
   Consider references to that shown in Figures 2.40 to 2.42, as well as to tropical cyclones.

Figure 2.39 The barograph trace of cyclone Elsie, recorded at Mandoora homestead.

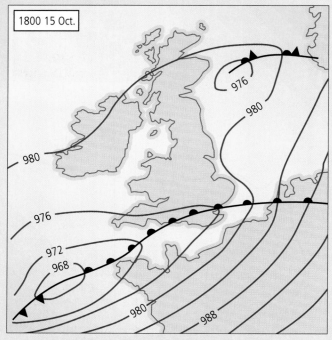

Figure 2.40 *The depression develops over the Bay of Biscay, with warm southerly winds over France.*

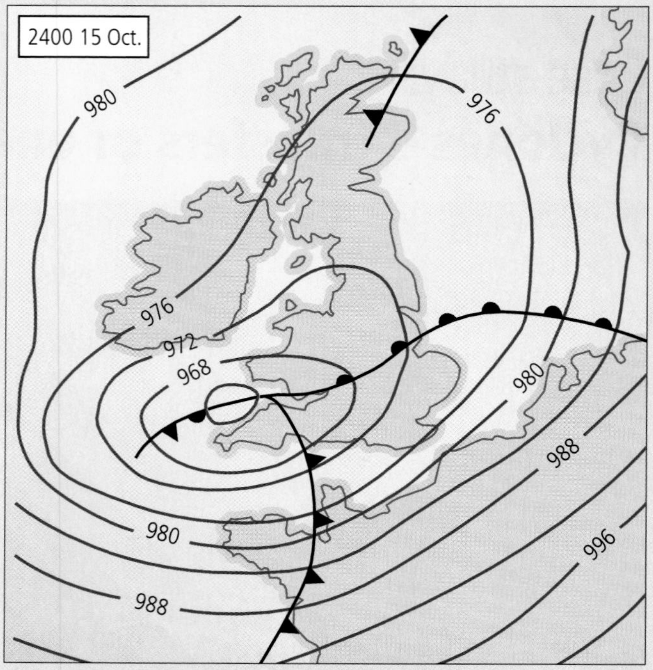

Figure 2.41 *The low deepens. Warm air rises rapidly, with cold air moving forward at the surface.*

Cyclonic development and the gale of October 1987

During the morning of 16 October 1987 an intense cyclonic system with exceptionally strong gales caused considerable damage across southern England. Its unusual path and the rapidity with which the low deepened were in response to great energy exchanges between near-surface air and the upper air.

On the evening of 15 October the low centred over the Bay of Biscay had begun to deepen. Strong winds from southern Spain began moving across the warm sector. A very strong upper-air gradient developed between a high over Spain and a low west of Ireland, with the geostrophic wind becoming a jet stream. Divergence in the northward-flowing jet caused upward air movement and convergence of surface air. This lead to the cyclonic system deepening even more rapidly, aided by instability over warm seas.

The cold front quickly narrowed the warm sector and caused occlusion. As the depression moved over the North Sea in the early morning of 16 October, the occluded front swung back around the centre of the depression (Figure 2.42). Around the centre, as indicated by the isobars, exceptionally strong, gusty winds damaged property and woodlands, particularly in south-eastern England.

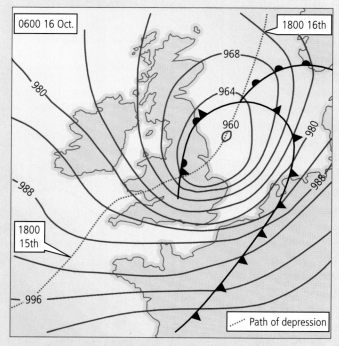

Figure 2.42 *Isobar spacing indicates strong winds over southern England as the occluded front swings towards the centre of the low (after Coones et al., 1988).*

Cyclonic systems over east-central USA

South-eastern USA experiences late summer hurricanes developed over the western Atlantic. But strong cyclonic disturbances also develop earlier. Energetic tornadoes, known as 'twisters', form and move rapidly over the countryside, their winds doing immense local damage before subsiding. They occur as warm, humid air is drawn over heated surfaces, which supply sufficient energy for the air above to become unstable and to rise extremely rapidly. With cool air sinking on one side, the air between begins to spin quite violently.

Figure 2.43 shows that during spring indrawn mT air creates energetic twisters in the southern parts, and that during summer these occur further north, with hot surfaces and humid air more frequent in higher latitudes. Regions with high annual incidence are indicated, but tornadoes may occur over most of the south-eastern parts of the country.

Apart from these, cyclonic systems bring heavy storms to the interior, often with thunder. These occur because of the great differences in temperature and humidity between air from inland and oceanic sources in spring and early summer. The synoptic charts show what weather contrasts can occur within a few weeks. In late spring, air of cP origin brings very cold conditions far to the south (Figure 2.44); while in early summer, very warm, humid mT air is drawn in, causing violent thunderstorms in the west, where cold and warm air clash, and in the far south favouring the development of 'twisters' (Figure 2.45).

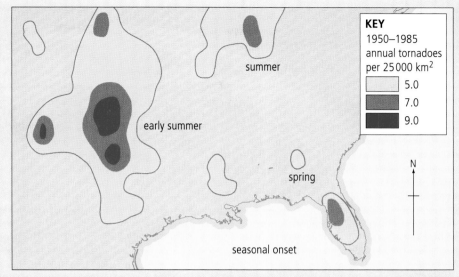

Figure 2.43 *Areas where the mean numbers of tornadoes* (twisters) *a year exceeds five; though during summer they develop over most of these eastern-central parts. Northward their onset tends to be later in the season.*

QUESTIONS

1 From Figures 2.40 to 2.42 describe the shifts in wind direction and strength over south-east England between 1800 and 0600 hours.

2 A report from southern England included – 'after light showers the temperature rose; then a heavy burst of rain; then a fall in temperature and winds up to gale force'. Relate this to the meteorological conditions.

3 Why should wind over the sea tend to have a higher speed than over land surfaces?

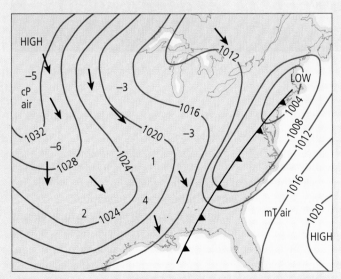

Figure 2.44 *A synoptic chart for late April. Surges of cPk air can bring very low temperatures to the far south.*

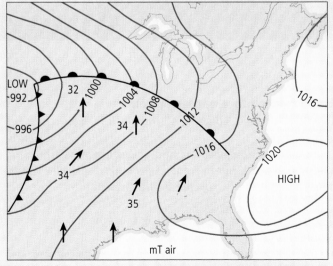

Figure 2.45 *During May hot humid air moves into eastern-central USA, with thunderstorms developing where it clashes with the colder air. Such conditions favour the development of tornadoes in southern parts.*

Figure 2.46 *A satellite view during June, with both clouds and lack of cloud indicating active weather systems and areas of air subsidence* (European Space Agency).

Cloud related to weather systems

Together Figures 2.46 and 2.47 provide a realistic summary of responses to Earth's zonal climatic variations and clearly show where circulating air and ocean currents transport energy poleward, but in some cases make for cooler conditions than might be expected for the latitude.

The clouds, or lack of clouds, shown on the satellite image (Figure 2.46) indicate vertical air movements and active weather systems for a day in June.

1   Clear skies point to air subsidence over the Sahara and Arabia.

2   Above the ocean west of the Sahara, skies are mostly clear, for cT air moves westward over the cold waters of the Canary Current.

3   The sky is also clear over the cold Benguela Current flowing northward along the desert/semi-desert lands of Namibia-Angola, though sea-fog or low cloud lies beneath subsiding air along their coastlands.

4   In the North Atlantic there are individual cumulus clouds in the mT air of the Trade Winds, but cloud masses form as the air becomes less stable air towards the Caribbean.

5   In the South Atlantic, lines of cumulus also travel towards the cloudy Brazilian coast, while west of the Benguela Current swirls of cloud mirror the ocean circulation.

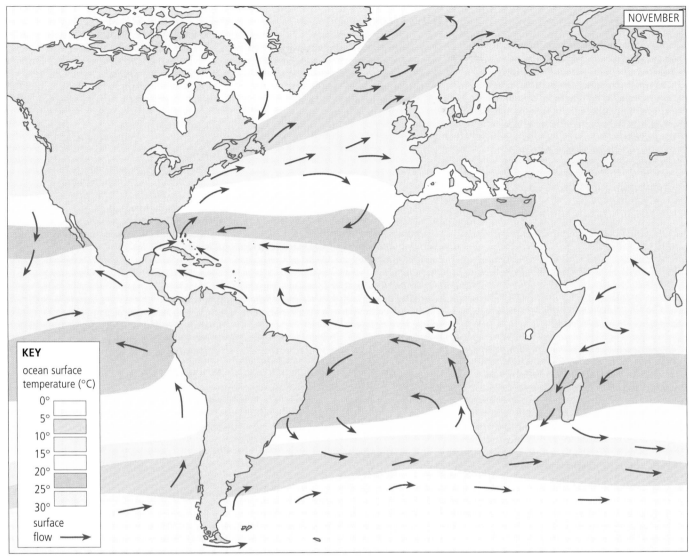

Figure 2.47 Ocean circulation and air temperature during November – the information from satellite radiance measurements (after Henderson Sellers and Robinson, 1999).

6   Near the Equator the Inter-Tropical Convergence Zone is clearly marked by the line of cloud across the ocean and by dense cloud masses forming over western Africa and extending into the Sahel.

7   In East Africa, a strong tropical cyclone moves along the south Kenyan coast.

8   Most of the Mediterranean is cloudless as cT air extends northward over western Europe, with the Alps marked by a line of cloud.

9   By contrast, an Atlantic depression is affecting Portugal and Spain, with another to the west of the British Isles.

10   Other lows appear to be deepening where air over the Gulf Stream, shown by a northward sweep of cloud, meets cold air from the Arctic.

11   Lines of cloud and cyclonic systems can also be seen in the temperate zone of westerlies across the South Atlantic.

Ocean circulation and surface temperature

Figure 2.47 shows ocean circulations and ocean surface temperature during November, highlighting the strong westerly flow in the South Atlantic and the passage of relatively warm water towards western Europe, continuing north-eastward into the Arctic. The movements of cold and warm offshore currents can be detected, affecting the air above and influencing adjoining coastal regions. Of course the heat storage capacity of the ocean is huge compared with that of the atmosphere, and the currents transport large amounts of heat energy.

QUESTIONS

1   Explain why Figure 4.27 shows a difference in temperature of offshore waters to the east and west of South Africa.

2   Where the Labrador Current (W of Greenland) and the Gulf Stream-North Atlantic Drift meet: (a) cyclonic lows develop; (b) there are productive fishing grounds. Suggest why.

SUMMARY

In this chapter we have seen how the circulation of air and ocean currents help to redistribute energy from lower to higher latitudes. The wind patterns and strength vary seasonally as changes in surface temperature affect the regional air pressure. The air's humidity, or lack of it, also makes for climatic variation, as in the summer and winter monsoons. The actual amount of precipitation received varies according to location, whether maritime or continental, and also depends on local relief.

Air temperature contrasts are greater in higher middle latitudes than within the tropics, and there frontal clashes between warm and cold surface air favour cyclonic development. Above, in the upper air, the poleward temperature gradient produces the strong looping westerlies with exceptionally fast jet streams. Their meanders create vertical air movements that also affect surface circulations and weather conditions. Thus in these latitudes local weather can change abruptly with the passage of an eastward-moving depression.

The sub-tropical zones, where air subsidence makes for surface aridity, are a key to climatic conditions far afield, for they feed the Trade Wind flows and supply cT and mT air to higher latitudes.

In the outer tropics, summer evaporation from heated ocean surfaces releases immense energy on condensation, which, with the Coriolis effect, may create the cyclonic swirls of a typhoon. In lower latitudes, where the Coriolis force is weak, convection from heated surfaces can still set up rainstorms.

The characteristics of the contrasting climatic regions described in Chapter 3 are closely related to the distribution of these patterns of energy, moisture and weather systems.

QUESTIONS

1 In temperate latitudes, why does rain gradually intensify as a warm front approaches, while heavy rain, the so-called 'clearing-up shower', signals the passage of a cold front?
2 What are the characteristics of northern Africa's air mass? How does its cT air affect seasonal weather conditions:
   a in Mediterranean lands?
   b in northern Nigeria?
3 Explain:
   a Why ITCZ describes a 'zone', rather than a 'front'.
   b Why its mean location varies seasonally.
   c Why its advance heralds the wetter season for many African countries.
4 Why do tropical easterlies tend to become more unstable as they cross the Atlantic, with more storm rainfall towards the west?
5 Explain the influence of 'air subsidence', 'a deepening thermal low' and 'high relief' on the rainfall distribution in the Indian sub-continent during the monsoons.

BIBLIOGRAPHY AND RECOMMENDED READING

Atkinson, B., 1998, *Modelling weather & climate*, Geography, 83(2), 147

Barry, R. and Chorley, R., 1998, *Atmosphere, Weather & Climate*, Methuen

Coones, P., Smith, G. and Burt, T., 1988, *The great gale of 16 October 1987*, Geography Review, 1(3), 6

Henderson-Sellers, A. and Robinson, P., 1999, *Contemporary Climatology*, Longman

Higgitt, D., 1998, *A drop in the ocean (currents)*, Geography Review, 11(3), 13

Miller, P., 1987, *Tracking tornadoes*, National Geographic, 171(6), 690

O'Hare, G., 1997, *The Indian monsoon*, Geography, 82(3), 218 and 82(4), 335

Salmond, J., 1994, *Hurricanes, Geography Review*, 8(1), 17

Smithson, P., 1993, *Tropical cyclones & their changing impact*, Geography, 78(2), 170

Strahler, A. H. and Strahler, A.N., 1992, *Modern Physical Geography*, Wiley & Sons

WEB SITES

European Centre for Medium-Range Forecasts – http://www.ecmwf.int/

TORRO: tornado and storms – http://www.zetnet.co.uk/oigs/torro

Tropical Storms Worldwide – http://www.solar.ifa.hawaii.edu/Tropical/tropical.html

World Meteorological Organisation (UN) – http://www.wmo.ch/

Chapter 3
The pattern of climates

In this chapter both descriptive and quantitative information indicate the part weather elements play in the world's climatic regions. But, as Earth's surfaces are far from uniform, the exact location and extent of a climatic region is difficult to define, and we find that within those regions described there are numerous variants.

DEFINING CLIMATIC REGIONS

How to identify similar regions

Defining climatic regions is bound to be hazardous and risk over-simplification. There is seldom a sharp climatic divide in nature and rarely, if ever, do all climatic elements in one location have exactly the same characteristics as in another.

We need to identify significant common features. But, as indicated, mean figures for rainfall or temperature can be misleading. In one area persistent high humidity, cloudiness and steady rain can give the same monthly rainfall figure as that for an area where heavy storms alternate with dry spells. The former may have small daily and monthly temperature ranges, the latter considerably larger ones.

Vegetation as a basis for classification

Regional climatic characteristics are usually reflected by the vegetation – though the present vegetation may not be a simple response to climate, but also to soils, poor drainage and interference by people, or fire.

Classifying climate on the basis of vegetation requires an appreciation of how during the year climatic elements affect the soil-water available for plant growth. The actual loss of water from soils and plant cover together is termed **evapo-transpiration** (Ea). The *potential* evapo-transpiration (Ep), the amount that would leave sufficient water to maintain the cover of vegetation, can be calculated.

Figure 3.1 shows, for each month, surpluses or deficits in the amount needed to maintain a particular plant cover – in this case in a humid mid-latitude location with cold winters and warm summers. P is the precipitation; R a surplus; D a deficit; −G a decrease by evapo-transpiration; and +G a recharge by precipitation.

The figure shows how natural vegetation might be affected and suggests that in a cultivated area the deficit D would have to be made up by irrigation. Figure 3.2 shows that this is so at Raipur. It has an east-central location on the Indian peninsular plateau. Here high temperatures during the short wet monsoon lead to water surpluses soon being

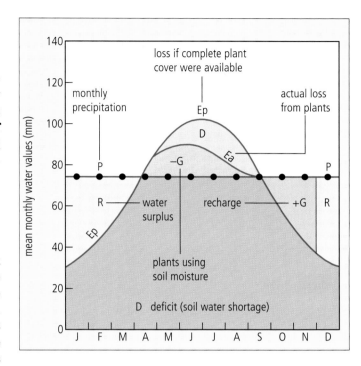

Figure 3.1 *A soil-water budget expressed in terms of the actual evapo-transpiration (Ea) and that required to maintain the plant cover (Ep). Here, for clarity, the precipitation (P) is shown as the same for each month.*

withdrawn, and with months of deficits the soils support forms of dry savanna. Such soil-water budgets, combined with figures for temperature regimes, can be used to identify regional similarities and lead to climatic classification. However, evapo-transpiration figures are not readily obtainable world-wide, so that other classifications, such as Köppen's, have used more available climatic statistics.

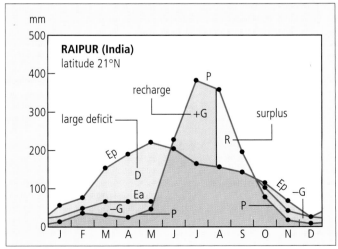

Figure 3.2 *The soil-water budget for Raipur on the tableland of the Indian peninsula (Strahler and Strahler, 1992).*

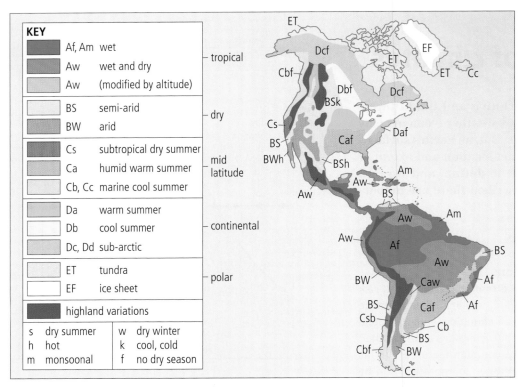

Figure 3.3 *The distribution of climates based on Trewartha's modification of Köppen's classification. Notice the contrasting regions in the middle latitudes to the west and east of both the Rockies and the Andes.*

In 1918 Dr W Köppen published a classification, which he and others later modified, based on conditions for growth required by groups of plants – by those favouring warm habitats to those thriving in colder environments. He chose mean values of temperature and precipitation to indicate **precipitation effectiveness** for plant growth, and located arid and semi-arid lands by the extent to which evaporation exceeded precipitation. For near-polar regions he considered conditions that limited tree growth, and for polar regions he considered temperatures that favoured at least one month with snow cover.

His five main groups were designated by letters *A* to *E*, and some groups were sub-divided, so that *BW* showed an *arid* region (*W* = Wüste, desert) and *BS* a *semi-arid* region. A third letter added further information: *h* = heiss, hot; *k* = kalt, cold; *m* = monsoonal; *f* = each month with over 60 mm precipitation; *w* = at least one month under 60 mm.

Table 3.1 *The main Groups*

A **Tropical Rainy:**	A hot climate with no cool season: the average temperature of each month is over 18°C
B **Dry:**	Evaporation exceeds precipitation
C **Humid Mesothermal:**	the warmest month has a mean temperature above 10°C; the coldest month has a mean temperature between −3°C and 18°C
D **Humid Microthermal:**	the warmest month has a mean temperature above 10°C; the coldest month has a mean temperature below −3°C
E **Polar:**	no month averages over 10°C

G.T. Trewartha's modification of Köppen's system (Figures 3.3 and 3.4) used the main groups and sub-groups, but as the original did not consider causes of the conditions described, his Table included references to the mode of origin of some of the climatic features (p. 106).

From 'climatic groups' to regional climates

In regional geography broad generalisations, as in the above groups and sub-groups, may not portray actual conditions within the designated region. Local studies within the main area will find unique climatic/vegetational conditions in mountainous territory or flood basins, or adjacent to another climatic group. This can be seen in south-western China, where, classified as part of a *Caw* climate, there are many localities with their own distinctive climate and numerous micro-climates.

Figure 3.5 shows the location of Dian Chi, a lake in central Yunnan large enough to modify the regional climate and allow lakeside Kunming to be known as 'city of eternal spring'. It particularly illustrates the difficulties in classifying climate on a large scale and the problems of doing this through climate-plant associations. For example, just to the south are tropical rainforests, yet to the north conifers dominate a great variety of vegetation on Yunnan's border with cloudy, mountainous south Sichuan ('Yunnan' in fact means 'south of the clouds'), while a short distance eastward are extensive, almost bare surfaces of dry limestone, broken into innumerable karst pinnacles. A comparison of the distribution of China's natural, but much modified, vegetation in Figure 3.5 and the extent of the *Caw/Caf* area in Figure 3.4 emphasises the limitations of broad classification based on climate-vegetation relationships.

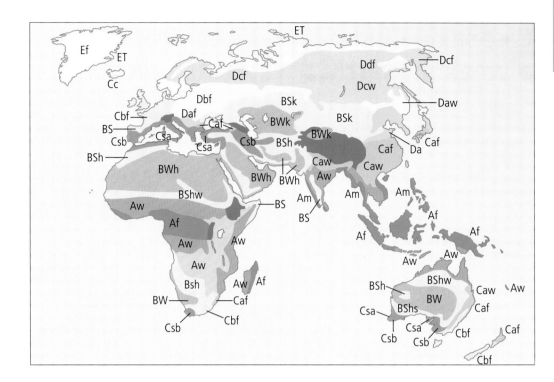

Figure 3.4 *Trewartha's modification of Köppen's classification. Notice how in southern Asia relief and rainshadow effects may give rise to* Aw *conditions rather than those of the hot wet monsoon* (Am)*.*

The following case study also emphasises that during a single week in China, in territory shown as being in these two main climatic groups, there is a wide variety of contrasting atmospheric conditions and associated forms of land-use. Be aware, therefore, that the broad summaries of global climates that follow, and the accompanying statistics and regional maps, mask many such variations.

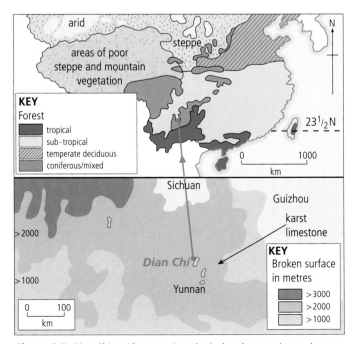

Figure 3.5 *Dian Chi amid contrasting physical and vegetation regions.*

Figure 3.6 *Dian Chi, the largest lake in south-west China.*

Case study: **China: climatic contrasts – a week in March**

Regional variation within climatic groupings, here *Caf-Caw*, is particularly illustrated by the responses of natural vegetation and of rural activities. In most countries seasonal activities are adapted both to broad climatic characteristics and to long-experienced changes in the weather – the likelihood of sudden frost, or storm-flooding, or whatever. Some regions are subject to sudden weather changes, others are more stable. This is amply illustrated in China by comparing regional activities during a single month.

It is therefore essential to start with an appreciation of scale – as the country extends for about 4000 km from north to south, and rather more from east to west. Overall it is affected by eastern Asia's winter and summer monsoons (p. 38), yet there are great contrasts in regional and local weather conditions and a corresponding variety of seasonal rural activities. Consider the different farming practices about villages in the loess lands (L), in the inland basin of Sichuan (S), and in tropical conditions in southern Yunnan (Y) during a single week in March – the transition period between the winter and summer monsoons – and how these relate to climatic influences.

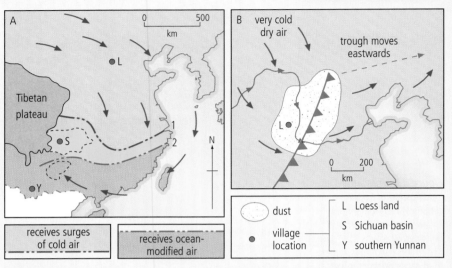

Figure 3.7 *Patterns of air movement during winter and early spring.* **A** *Towards the south semi-permanent 'fronts' (1 and 2) develop in the zone between the cold dry northerlies and the warmer, moister air.* **B** *A trough within the very cold dry air moves eastward, stirring up dust.*

During winter there are surges of dry, bitterly cold air from the high pressure over Siberia and Mongolia. As troughs cross northern and central China (Figure 3.7), strong cold winds raise great clouds of loose loess from the dry surfaces, which are carried as dust over north-east China as cold fronts move eastward. Figure 3.8 shows that at L, even as late as March, the terraced village fields above the dry, icy river course are bare. Nearby, sheltered by a loess terrace wall, men are ploughing in preparation for the spring wheat.

As air circulating about these western high pressure centres passes out over the ocean and turns southward, it becomes warmer, moister and less stable. Shallow flows of this modified cP air swing into southern China. Also as summer approaches, the far south receives incursions of mT air. Between, in south-central China, where cold surges from the north meet warmer moist inflows, weak cyclones develop, bringing occasional storms. However, the western basin of Sichuan is mostly sheltered by mountains from the cold northerlies and avoids cyclonic

Figure 3.8 *In early spring the river valley, cut into thick loess, is icy; the terraced fields below the village are not yet planted.*

Figure 3.9 *Ploughing a loess terrace in readiness for spring wheat; the trees are still bare of leaves. Beyond, the cohesive property of the loess is being used to make bricks.*

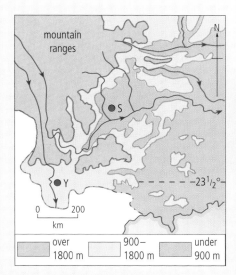

Figure 3.10 *The climate in the south-west is strongly influenced by the western highlands. In winter the Sichuan basin benefits from the shelter of surrounding ranges.*

development. With air from the western heights sinking and warming, its winters are mild, though moist air from the south-east makes for cloudiness. This is an equable fertile part of the country, with intricate canal systems from rivers rising in the mountains. In March inputs of solar energy are increasing and at S the winter wheat is being harvested and adjacent fields flooded and planted with seedlings for the summer rice crop (Figure 3.11).

Further south, in Yunnan there are mountains, plateaux and deeply incised rivers, but also fertile upland basins and broader valleys with a tropical all-year-round growing season. Winters are still the drier season, but before the main inflow of warm humid air from the south-east, lows related to disturbances in the sub-tropical jet stream bring rain from the south-west. At Y during March, men from a village set amid the leafy trees and tall bamboo are ploughing for their second rice crop.

The contrast in environments and activities around each of these villages shows how little is revealed about actual climatic characteristics by describing east Asian climates simply as 'monsoonal' and stresses the need to appreciate regional and local influences.

Figure 3.11 *March in the south of the 'red basin' of Sichuan, where the wheat is being harvested while villagers are planting out rice seedlings in flooded padi fields.*

QUESTIONS

1 Suggest why climatic characteristics in various parts of China appear to support the observation that 'sub-tropical areas between latitudes 23° and 40° experience wide variations of both temperature and rainfall, whereas within the tropics, where temperatures remain relatively constant, rainfall variations dominate the patterns of climate.'

2 Figure 3.7A shows the mean positions in March of a semi-permanent 'cold front' (1) and a 'warm front' (2), though in fact their advances vary from time to time. Why, at times, do clouds form and small depressions develop within the zone between?

3 Suggest why the Sichuan basin, though sheltered, is likely to be cloudier than central Yunnan. Consider why each gains warmth during winter from katabatic winds.

Figure 3.12 *March in Yunnan, with bamboos and trees in leaf. Men in light shirts are using buffaloes to plough flooded fields for their second rice crop.*

LOW LATITUDE CLIMATES

Tropical rainy (*Af*)

This is experienced by large areas of relatively low land within 5° to 10° of the Equator, but extends to those near-coastal regions in higher latitudes that receive rain from unstable tropical easterlies (Figure 3.13).

The noon sun is never far from the zenith, though high humidity and cloud prevent the temperature soaring. There is little seasonal variation, with mean monthly temperatures some 25°–27°C and a daily range of about 8°–10°C. The mean annual rainfall is high, though in response to the movements of the ITCZ some months are relatively wetter or drier. The humidity remains high, even though skies tend to clear at night. During morning cumulus develops, and midday heat frequently creates towering cumulonimbus, with heavy thundery downpours. Weak cyclonic systems do develop, and some days have more cloud, some less.

Figures 3.14 to 3.17 show how rainforest vegetation responds to all-year-round heat and rainfall, which allows nutrients from plant debris to be recycled. Clearance disturbs these balanced processes, but on suitable soil-material stable agriculture can be most productive (Figure 3.14).

Figure 3.14 On fertile soils tropical rainforest may be cleared for cultivation of all-year-round grain and tree crops, as here on Indonesian volcanics.

Table 3.2

Uaupes, Amazon Lowlands. 0° 08′ S. 83m														Max/min °C	
	J	F	M	A	M	J	J	A	S	O	N	D	Total	Year	Absolute
°C	26	27	26	26	26	26	25	26	27	27	27	26	—	31	38
mm	262	196	254	269	305	234	223	183	132	175	183	264	**2677**	22	11

Singapore, South-East Asia. 1° 18′ N. 10m														Max/min °C	
	J	F	M	A	M	J	J	A	S	O	N	D	Total	Year	Absolute
°C	26	27	27	27	28	27	27	27	27	27	27	27	—	31	36
mm	252	172	193	188	172	172	170	196	178	208	254	257	**2413**	23	19

Figure 3.13

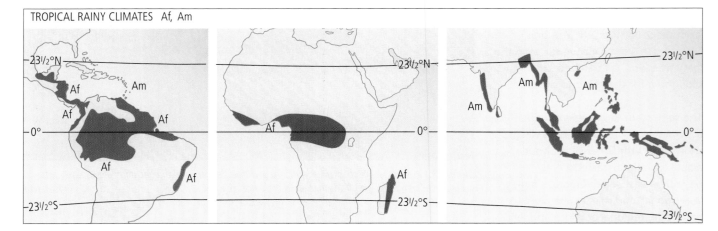

TROPICAL RAINY CLIMATES Af, Am

Table 3.3

Bombay, Western India. 18° 54′ N. 11m															Max/min °C		
	J	F	M	A	M	J	J	A	S	O	N	D	Total		Jan	May	Absolute
°C	23	23	26	28	30	29	27	27	27	28	27	26	—		28	33	Jan Mar
mm	2	2	2	2	18	486	618	340	264	64	13	2	**1808**		19	27	12 38

Chittagong, Bangladesh. 22° 21′ N. 26m															Max/min °C		
	J	F	M	A	M	J	J	A	S	O	N	D	Total		Jan	Jun	Absolute
°C	19	21	25	27	28	28	27	27	27	27	23	20	—		26	31	Jan Mar
mm	5	26	64	150	264	533	597	518	320	181	56	15	**2831**		13	25	7 39

Tropical wet monsoon (*Am*)

During the dry winter the mean temperatures in these regions remain relatively high, rising to a maximum just before the 'burst' of the wet monsoon. However, they do fall somewhat during this cloudy wet period, when the inflows of mT air bring exceptionally high rainfall to windward slopes. The wetter locations support deciduous forest; but though the drier leeward areas are monsoonal in their regime, their vegetation is mostly tree-grassland, resembling that of the *Aw* regions.

Malaysia and Indonesia are sometimes seen as monsoonal sub-types of the *Af* climate. This is because from May to September there are inflows of air into Asia, and then from December to March air moves towards northern Australia, bringing rain at times throughout the year.

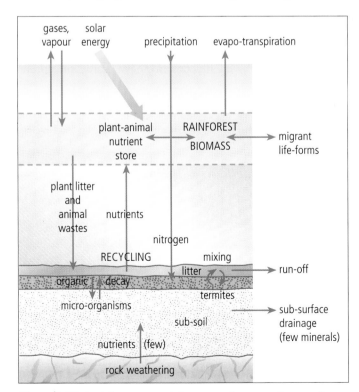

Figure 3.15 A rainforest system thrives on continuous nutrient recycling. All-year-round inputs of energy and water maintain this finely balanced exchange.

Figure 3.16 The majority of rainforest species need access to solar energy, and here a tall emergent is festooned with vines, lianes and epiphytes, each with a different method of obtaining sufficient energy for growth.

Figure 3.17 Brazilian rainforest, where nutrients are cycled between the abundant vegetation and this shallow soil layer. Below is a deep red infertile regolith on weathered granite.

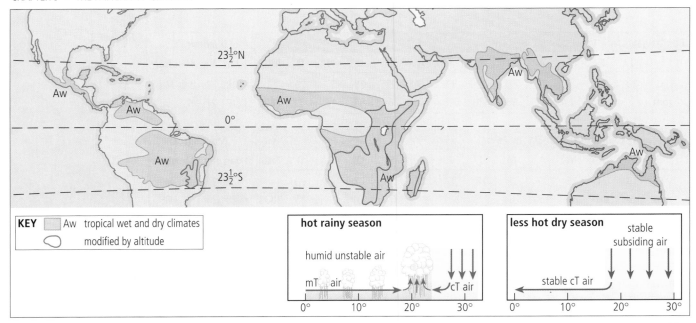

KEY Aw tropical wet and dry climates
   modified by altitude

hot rainy season

humid unstable air

mT air   cT air

0° 10° 20° 30°

less hot dry season   stable subsiding air

stable cT air

0° 10° 20° 30°

Figure 3.18 *There are considerable climatic variations in these Aw regions. Apart from contrasts in altitude, rainfall responds to seasonal movements of the ITCZ and varies with latitude.*

Tropical wet season/dry season (*Aw*)

Poleward of the *Af* climates there tends to be a gradual transition to an *Aw* climate, the dry season becoming longer with increasing latitude. Rainfall amounts and the type of vegetation thus vary considerably from the moist low latitudes to the semi-desert margins. During the less hot season, with more stable air, the rainfall is low (Figure 3.18). However, convection can still create swirling dust-devils in drier parts.

As the noon sun nears the zenith, temperatures increase, air pressure falls and strong convection makes for thundery storms in the moist air. As the angle of the noon sun decreases, once more the rains gradually cease and drier air is re-established.

Figure 3.18 shows the distribution of *Aw* climate, though most of these lands contain plateaux and highlands, which modify temperatures. Variations in soils and micro-climates abound within the vast areas broadly described as 'savanna lands'. Savanna vegetation has dominant tropical grasses, but varies from coarse tussocks with thorn scrub in dry areas, through parkland savanna with scattered trees and bushes, to close woodland with tall grass in moister locations. However, it is unwise to attribute a particular form of savanna solely to climate, and it is inaccurate to use the term 'savanna climate' – for throughout these lands widespread clearance, animal grazing and frequent fires affect the tree-grassland associations.

Figure 3.19 *The dry season on the Tanzanian savanna, where herdsmen drive cattle to water in the course of a shrunken stream, which is capable during the rains of eroding this steep cliff face.*

Table 3.4

Sokoto, Northern Nigeria. 13° 01' N. 350m														Max/min °C		
	J	F	M	A	M	J	J	A	S	O	N	D	Total	Dec Aug	Absolute	
°C	24	26	31	33	33	30	28	26	27	29	27	25	—	33 41	Jan Apr	
mm	—	—	2	10	43	94	152	244	132	13	—	—	691	16 26	7 47	

Goias, Central Brazilian Plateau. 15° 58' S. 536m														Max/min °C		
	J	F	M	A	M	J	J	A	S	O	N	D	Total	Jun Sep	Absolute	
°C	23	24	24	25	24	22	22	24	26	26	24	23	—	32 34	Jun Sep	
mm	318	252	259	117	10	8	—	8	59	135	239	242	1646	13 18	5 40	

Figure 3.20 *Tall grasses and trees in leaf during the rains stress the variety of vegetation within the East African* Aw *climate. Here, even in the dry season, humid air from nearby Lake Manyara helps to maintain close tree savanna.*

Precarious conditions of this kind also occur in north-central and western Australia, south-west Africa and parts of the dry enclave of eastern Brazil.

QUESTIONS

1 Explain how the landscape features in Figure 3.19 support the fact that most *Aw* regions are a mosaic of soils and micro-climates with varying vegetation, even over small areas.

2 Describe and account for the variety of physical features within the valley itself, and stress how they relate to the broad characteristics of an *Aw* climate.

Semi-arid outer tropical (*BShw*)

Bordering the hot deserts, these lands have long, dry winters dominated by air subsidence. Brief, erratic summer rains occur with the ITCZ near its poleward limit, but during these hottest months, with rapid evaporation, it less effective for plant growth. As in the Sahel zone, south of the Sahara, years with average rainfall may be followed by many years of drought (p. 87).

Figure 3.21 *Vast areas with spinifex, salt-bush, blue-bush and small herbs border the arid plateau lands of central and western Australia.*

Table 3.5

Hall's Creek, North-West Australia. 18° 13' S. 366m														Max/min °C		
	J	F	M	A	M	J	J	A	S	O	N	D	Total	Jul Nov	Absolute	
°C	30	29	28	26	21	19	18	21	24	28	31	31	—	27 38	Jul Jan	
mm	137	107	71	13	5	5	5	2	2	13	35	79	474	9 23	−1 44	

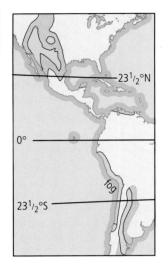

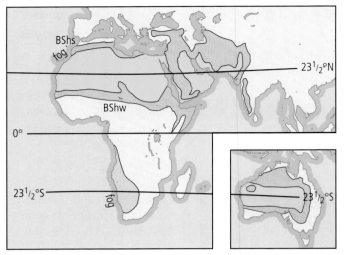

Hot desert and semi-arid climates

KEY		
	BWh	hot desert
	BShs BShw	semi-arid

Figure 3.22 *The extent of the semi-arid fringes of the hot deserts are difficult to define climatically, and their vegetation fluctuates considerably as wet periods alternate with lengthy droughts.*

Hot desert (*BWh*)

In the sub-tropics subsiding air affects these very arid western parts of the landmasses between latitudes 20° and 25° and strongly influences adjacent lands. Even though during summer the air contains a considerable amount of water vapour, with intense heat the relative humidity is low. Stable subsiding air suppresses convective updraughts, which seldom reach sufficient height for cumulonimbus to develop, though occasional downpours do cause sheet-wash and flash floods, which soak into depressions.

While by day the summer shade temperature may reach 50°–55°C, at night, with energy lost through clear skies, it can fall to some 20°–24°C. During winter, with day temperatures of 15°–20°C, near-surface night temperatures are often low enough to allow heavy dews to form.

Semi-arid: poleward of hot deserts (*BShs*)

Summer months are dry and very hot (*s* = *summer drought*). During winter occasional rain associated with mid-latitude depressions affects these semi-arid regions. But the extent of these marginal lands is difficult to define on a rainfall basis, for years of drought may be followed by storms bringing hundreds of millimetres of rain in an hour or so. Generally the winter rain supports coarse grasses and drought-tolerant plants.

Such conditions occur over large tracts of north Africa, northern Arabia eastward to north-west India and also in south-central Australia. In semi-arid regions of northern Mexico and adjoining parts of USA, a summer maximum rainfall is more common.

Figure 3.23 *Here in arid Sinai what little rain falls comes during winter. Plant life is supported mainly by ground-water, replenished by drainage into the valleys. A few deep-rooted trees survive amid the debris left by an occasional flash-flood.*

Table 3.6

In Salah, Southern Algeria. 27° 12′ N. 280m														Max/min °C			
	J	F	M	A	M	J	J	A	S	O	N	D	Total	Jan	Jul	Absolute	
°C	13	16	20	25	29	35	37	36	33	27	19	14	—	21	45	Jan	Jul
mm	2	3	*	*	*	*	—	2	*	*	3	5	**15**	6	28	−3	50

* - Less than 2mm

Case study:
The Andes – climatic diversity

Figure 3.24 *The structure of the Peruvian Andes. Vegetation reflects east–west climatic contrasts.*

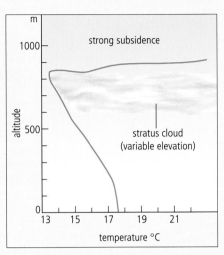

Figure 3.25 *Climatic influences east and west of the Andes during December.*

Figure 3.26 *Inversion conditions aloft making for retention of moisture and pollutants in the lower air.*

From *BWh* desert to *Af* rainforest

Each of the great mountain chains, the Himalayas, the Rocky Mountains and the Andes, with their differences in altitude and aspect, exposure and shelter, not only contain numerous locations with an individual climate, but also are bordered by contrasting climatic regions. Thus the arid western rim of the Peruvian Andes rises above a desert coastland, whereas far to the east, beyond the raised tablelands, the high basins and volcanic peaks, are deeply gullied slopes and steep tree-covered canyon sides that merge at lower levels with the tropical forests of the Amazon basin (Figure 3.28).

Along the western shores strong subsidence, related to the sub-tropical high of the eastern Pacific, makes this one of the world's most arid deserts (Figure 3.27). The potentially stable air, circulating anti-clockwise around the high, is deflected northwards by the wall of the Andes, so that desert conditions extend far into the low latitudes.

Upwelling cold waters of the Peru current keep the air temperature below that of most sub-tropical deserts; while air subsidence is apt to cause mist or fog to develop both offshore and over the narrow coastland. Inversion conditions also tend to create low stratus cloud several hundred metres above sea-level – a source of moisture which establishes a low vegetation, including grasses, high up on the otherwise arid hillsides.

As Figure 3.29 shows, eastward among the mountains air humidity increases. At first the tendency for air to subside forms layer cloud above the high surfaces, disturbed in places by the rising air currents of mountain or valley winds. But over the eastern Andes towering clouds form in the unstable air, and abundant rain supplies high tributaries to the Amazon, which flow in deep ravines through the dissected foothills.

Figure 3.27 *Air subsidence maintains the aridity of Peru's coastal mountains. A few streams reach the coast and allow settlement based on irrigation.*

Figure 3.28 *Billowing cloud high above dissected, wooded mountainsides in the eastern Andes.*

Such climatic diversity was emphasised by the meteorological conditions and their influence on the disastrous events that occurred over the Christmas period of 1971. On Christmas Eve passenger planes due to fly eastward were held at Lima's airport. The local air was stable enough to trap a typical layer of mist and pollutants beneath grey strata (Figure 3.31), but there were reports that wave disturbances in unstable air over the western Andes were causing strong uplift. These were borne out, sadly, when the only plane allowed to take-off, for Pucallpa, was torn apart in a violent storm beyond the eastern foothills. The sole survivor, a young girl, managed to stay alive in the rainforest until rescued a month later. Torrential rains over the next few days caused landslides to cascade down the eastern mountain slopes and, in the narrow Urubamba valley derailed a passenger train on the single-track railway above the river (Figure 3.33). Such a period of particular instability occurs over the years in the moist tropical air east of the Andes, and in December 1999 caused devastating hillside slumps in Venezuela.

All of which shows that a blanket description of 'the climate' of a region can mask a variety of weather conditions that may vitally affect local peoples. This is particularly so among mountains; though here it is also true to the west of the Andes, where during the periodic El Niño events (p. 88) the coastal climate is drastically disturbed.

Diversity within the mountains

The assertion that mountains 'make their own climates' is borne out both by the east–west contrasts in

Figure 3.30 *Cotopaxi volcano in the eastern Andes, rising above the mists and clouds of the eastern foothills.*

atmospheric conditions and by the variety of climates within the high ranges. Figure 3.24 shows the trends of mountain chains, between which are numerous extensive valleys and high basins, many of which have fertile soils and sufficient rainfall to encourage arable farming. This is particularly so towards the east where the humidity is higher and snow-melt waters supply farming communities situated around sizeable settlements. For some farms vulcanicity associated with Andean uplift provides fertile soils. Other farms work on morainic and outwash material from formerly glaciated valleys, as in Figure 3.34.

Figure 3.29 *Across the dry western Andes to where cloud layers show that air subsidence is still checking buoyancy, while further east cumulus towers above the snowy volcanic peaks and clouds rise high above the foothills bordering Amazonia.*

Figure 3.31 *Subsiding air causes mist, smoke and other industrial and household pollutants to blanket Lima's suburbs about the Rimac river, whose flow is almost completely reduced by urban demands.*

At these heights energy exchanges between mountain surfaces and the atmosphere differ from those at low level (Figure 3.32). The natural vegetation and crops in the high valleys receive a greater intensity of insolation. Solar energy reflected from the more exposed mountainsides with relatively little plant cover also passes readily through the less dense atmosphere.

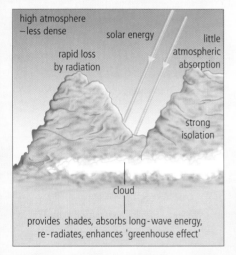

Figure 3.32 *In these latitudes high mountains and valleys receive strong insolation, and mostly radiate energy into a clear, thin atmosphere.*

Clouds, as already indicated, act to modify local climates to the east rather than to the west of the Andes.

As always, the nature of the surface affects the air above. For instance, amid the semi-arid plateau of southern Peru and Bolivia the wide waters of Lake Titicaca absorb and retain solar energy and add humidity to the air above, encouraging settlement on its islands and about the lake itself.

Figure 3.33 *Clearing the railway along the Urubamba valley of boulders from a talus slide following exceptionally heavy rain.*

QUESTIONS

1 When considering the variety of climates in and around the Andean chain, why is the stability or instability of the lower air such a major factor?

2 Describe and account for the topography of the arid coastal mountains seen in Figure 3.27. Which features indicate former climatic conditions, and what were they like?

3 The eastern mountain slopes are prone to landslides. Consider the ways in which freeze–thaw, periods of heavy rainfall, progressive erosion by rivers in ravines and railway construction along such river courses make for surface instability. Remember that this is also a tectonically active zone with many volcanoes.

4 Figure 3.24 shows some of the larger valleys, enclosed basins and drainage patterns high amid the mountain chains. Point out the variety of climatic and physical conditions that allow many of them to support a considerable population.

5 a Why does temperature inversion act to promote vegetation several hundred metres above the arid coastland?

   b How does such inversion affect the atmospheric conditions experienced in Lima (Figure 3.31)?

6 This case study examines climatic diversity related to the Andes. Find examples from other mountainous regions of variation of climate with altitude and of the consequent differences in vegetation and land-use.

Figure 3.34 *A glaciated valley with truncated spurs among the high mountain chains in central Peru, where an enclave of fertile farmland on morainic and outwash materials also benefits from strong sunlight.*

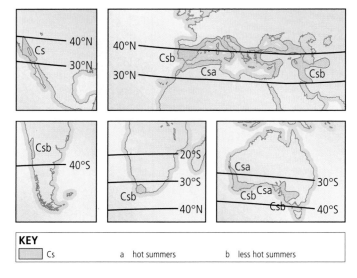

KEY

▢ Cs	a hot summers	b less hot summers

Figure 3.35 *The so-called 'Mediterranean' climatic regions, between latitude 30° and 40° to the west of the continents.*

MIDDLE LATITUDE CLIMATES

Warm temperate summer dry: Mediterranean type (Cs)

This is experienced on the western side of continents at about 35° latitude. Under the influence of dry subsiding air the summer is hot, with abundant sunshine and little rain; in winter, incursions of westerly air, in most cases over a cold offshore current, brings milder, moister conditions.

After dry, sunny summers the Mediterranean itself experiences winter incursions of humid air. There is frontal rain as occasional polar front depressions quickly move eastward, but followed by long sunny periods, ·favouring plant growth. Though winters are generally mild, night temperature may fall below freezing and the variable weather brings cold spells. Throughout the Mediterranean air movements about such cyclonic systems can cause cold air from northern mountains to flow southward as freezing winds like the Mistral and Bora. At times, however, they draw hotter, dry, dusty air from the southern deserts – the Khamsin and Sirocco winds.

There were once extensive mixed evergreen woodlands about the Mediterranean Sea that were adapted to the mild growing season and summer drought. After long human occupation much of the vegetation is now shrub-heath, with low plants adapted to the summer deficit by leathery leaves and an ability to close their stomata during drought – plant communities known as maquis and garrigue. In southern Europe the distribution of the olive, introduced long ago from further east, more or less indicates the extent of this *Cs* climate.

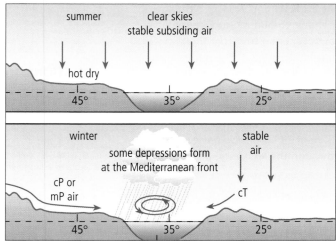

Figure 3.36 *Pressure systems affecting the 'winter wet–summer dry' regime of the Mediterranean lands.*

QUESTIONS

1 Which appears to be the most favourable part of the year for plant growth?

2 Quote figures from Table 3.7 that suggest there is a long growing season.

3 What indicates that the dry months favour the region's intensive irrigated farming?

Table 3.7

Nicosia, Cyprus. 35° 09′ N. 218m														Max/min °C		
	J	**F**	**M**	**A**	**M**	**J**	**J**	**A**	**S**	**O**	**N**	**D**	**Total**	**Jan Jul**	**Absolute**	
°C	10	10	12	17	22	26	28	28	26	21	16	12	—	14 36	Feb Jul	
mm	74	51	33	20	27	10	*	*	5	23	43	76	**364**	6 21	−5 47	

* - Less than 2.5mm

Lisbon, Portugal. 38° 43′ N. 95m														Max/min °C		
	J	**F**	**M**	**A**	**M**	**J**	**J**	**A**	**S**	**O**	**N**	**D**	**Total**	**Jan Aug**	**Absolute**	
°C	11	11	13	14	17	20	22	22	21	17	14	11	—	13 27	Feb Jul	
mm	84	81	79	61	43	18	5	5	35	79	107	91	**686**	8 18	−2 39	

Figure 3.37 *In the Mediterranean the mild wet winters and dry summers favour limestone weathering and the retention of its impurities. These, reddened by iron compounds, make up this terra rossa.*

Table 3.8

Cs regions			*Max/min °C		Total
Coastal	Inland		Summer	Winter	mm
San Francisco	38°N	16m	26/10	17/4	510
Fresno	*37°N*	*101m*	*42/14*	*20/−3*	*231*
Valparaiso	33°N	41m	27/11	20/5	506
Santiago	*33°N*	*510m*	*33/9*	*21/−2*	*358*
Italics - inland station					* - mean figures for Jan/Jul

Table 3.7 indicates differences between the marine-influenced, cooler and moister climatic sub-type *Csb* and the more extreme *Csa*. Table 3.8 shows that in other *Cs* regions there are contrasts in temperature and rainfall between coasts and the interior. The effects of cold offshore currents are common to each. In California fog forms offshore at times sweeping into the Bay (Figures 4.37 and 4.38). Similarly amid Cape Town's summer sunshine a westerly can bring a short, cool spell with light rain, and fog is frequently formed off Cape Point (Figure 1.19).

Figure 3.38 *Garrigue on a Cyprus hillside. The deep-rooted scented shrubs, broom, and conifers are adapted to survive the hot dry Mediterranean summers.*

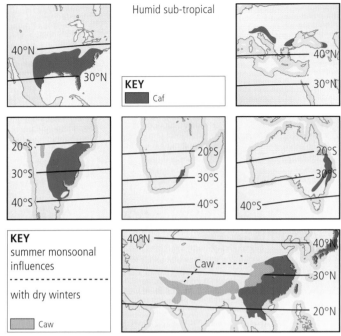

Humid sub-tropical

KEY

| | Caf |

KEY
summer monsoonal
influences
- - - - - - - - - - - - - - - -
with dry winters

| | Caw |

Caw - - - - - -

Figure 3.39 *For each region the characteristics of its continental hinterland influences the climate, making for variety among the Ca regimes.*

Humid sub-tropical (*Ca*)

In the eastern parts of continents in the lower middle latitudes and the sub-tropics there is no distinct dry season – a *Caf* climate. Tropical maritime air moves from the western parts of the oceans over the land areas, especially during summer when the land interior is hot. Though rain falls through the year there is a summer maximum, with thundery downpours from humid air with a mean temperature of 22°–30°C. Hurricanes tend to affect the eastern seaboard, especially in early autumn. Conditions become drier towards the interior.

In winter there is greater stability, when incursions of maritime air pass over warm, rather than hot, surfaces. In Australia and Argentina, winters are mild, with occasional frontal rain belts. However, in China, air from the Asiatic interior brings much lower temperatures. While in south-eastern USA, where occasional surges of cold air meet the maritime air, periods of frontal rain give some places a slight winter maximum rainfall.

In a sense the climate of each region is dominated by an influx of mT air from the western parts of the oceanic high pressure zones. Each of the maps in Figure 3.42 shows the mean summer location of these high pressure centres. Over warm oceans the encircling air becomes potentially

Table 3.9

Port Macquarie, Eastern New South Wales. 31° 38′ S. 19m														**Max/min °C**		
	J	F	M	A	M	J	J	A	S	O	N	D	Total	Jul	Jan	Absolute
°C	22	22	21	18	16	13	12	13	15	17	19	21	—	18	26	Jun Feb
mm	140	178	162	165	142	120	110	84	96	89	94	127	**1507**	7	18	−1 41

Charleston, South Carolina, USA. 32° 47′ N. 3m														**Max/min °C**		
	J	F	M	A	M	J	J	A	S	O	N	D	Total	Jan	Jul	Absolute
°C	10	11	14	18	23	26	27	27	25	20	14	11	—	14	31	Feb Jul
mm	74	84	86	71	81	120	186	167	130	81	59	71	**1205**	6	24	−14 40

Figure 3.40 *Sugarcane on cleared coastal woodland in the humid Caf climate of Natal.*

Figure 3.41 A village amid tall bamboos beside the Li jiang in south-central China. A Ca region, the karst hills closely wooded, but the winter sky cloudy, the people well clad.

Table 3.10

Nanjing, Eastern China. 32° 03′ N. 16m													Max/min °C		
	J	F	M	A	M	J	J	A	S	O	N	D	Total	Jan Jul	Absolute
°C	2	4	9	14	20	24	28	27	22	18	11	4	—	6 31	Jan Jul
mm	40	51	76	102	81	183	206	117	94	51	40	30	**1072**	−2 24	−13 40

unstable, so that storms develop over the heated land surfaces – resulting in the high summer rainfall (Tables 3. 9 and 3.10).

However, there is considerable regional variety in the *Ca* climates. Small parts of southern Europe have *Caf* charac-

teristics, such as the plains of northern Italy and the lower Danube basin, with precipitation through the year but a summer maximum. On the other hand, sheltered parts of eastern China have a *Caw* climate with very dry winters.

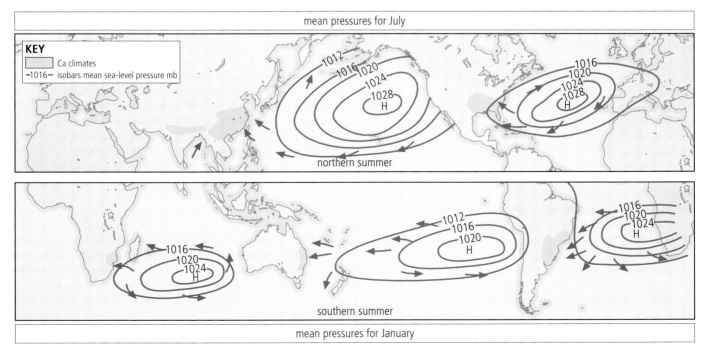

Figure 3.42 Caf *regions receive mT air, unstable after passage from high pressure sources over the oceans.*

Figure 3.43 *A semi-arid landscape in China's north-west, with fields on terraces of thick loess – deeply eroded accumulations of fine dust from the interior.*

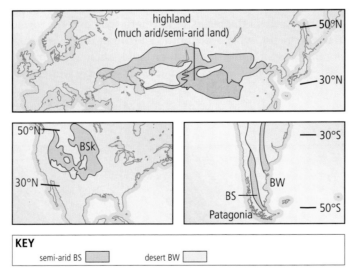

Figure 3.44 *Dry continental interior regions, their winter climate dominated by thermal high pressure. Patagonia is a special case.*

Middle latitude semi-arid (BSk)

These extensive semi-arid lands in central Asia and North America have extremely cold winters, so snow lies for long periods. Annual precipitation of some 200–300 mm is unreliable and varies with location. Moist summer inflows bring brief thundery storms, but evaporation exceeds precipitation. In the more arid locations their dry grasslands give way to patchy spiny-grass scrub, and in each continent extensive grazing and unirrigated agriculture have caused widespread erosion.

In Patagonia mean winter temperatures are above freezing. However, semi-arid conditions extend eastward in the rain-shadow of the Andes, and strong winds dry the tussocky grassland. Offshore the cold Falkland current helps to maintain coastal aridity.

Middle latitude arid interior (BWk)

In the heart of the northern continents, especially in Asia, precipitation is slight and irregular, mostly from summer convection of indrawn air. Winters are bitterly cold, so that, with humidity as low as 20%, snowfall is usually light, but remains on the surface for a long time. Air tends to sink into inland basins, which become extremely arid. But generalisations can be misleading, for such conditions occur

Table 3.11

Ashkhabad, Turkmenistan. 37° 57′ N. 226m														Max/min °C		
	J	F	M	A	M	J	J	A	S	O	N	D	Total	Jan	Jul	Absolute
°C	−1	7	8	15	19	26	29	24	22	15	8	4	—	3	36	Jan Jul
mm	25	20	48	35	30	8	3	3	3	13	20	18	**226**	−4	22	−26 45

Figure 3.45 *The moist cool-temperate western coastland of New Zealand's South Island, backed by the Southern Alps, with evergreen beech in response to mild westerlies that bring rain throughout the year. Introduced lupins have dispersed naturally from local homesteads.*

through some 20° of latitude. Also topography varies from high plateaux with outstanding ranges to low inland basins; while towards the west, where the winter high is weaker, maritime influences bring occasional rain or snow.

Cool temperate humid (maritime) (*Cb, Cc*)

About latitudes 40°–60° prevailing westerlies with eastward-moving depressions bring maritime influences over western parts of the continents. The extent of their penetration depends largely on relief. In western Europe, with extensive lowland, their influence is felt far inland, whereas in Canada and Chile high mountains restrict this type of climate to the narrow western coastlands. The opposite is also true, in that continental influences have greater effects on the climates of the western European countries.

Over western Europe a procession of eastward-moving cyclonic systems, with associated fronts, ridges and troughs, makes for changeable weather. Maritime influences keep mean summer temperatures to about 13°–18°C, depending on latitude and distance from the ocean. Lengthy droughts are rare, though extensions of sub-tropical high pressure occasionally create unusually dry spells in summer – while the invasion of continental high pressure during winter can cause periods of frost or persistent fog.

There are, however, marked climatic contrasts between western and eastern locations. Winter air from the relatively warm Atlantic keeps coastlands mild. Norwegian ports, even those in the Arctic, have mean temperatures some 15C° above the average for the latitude and remain open for shipping. But cold cP air does cover western Europe

Cool temperate humid climates Cbf and Cc

West Coast Maritime – rain in all seasons

Southern hemisphere – rain in all seasons

I cool winters   II warm winters

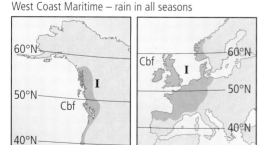
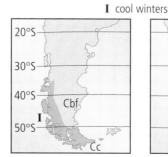
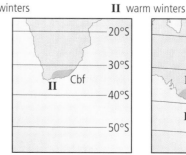
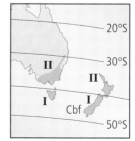

Figure 3.46 *Temperate **Cbf** regions with rain in all seasons. Most have cool winters (I); though South Africa, eastern mainland Australia and New Zealand's North Island have much warmer ones (II).*

Table 3.12

Shannon, Western Ireland. 52° 41' N. 2m														Max/min °C		
	J	F	M	A	M	J	J	A	S	O	N	D	Total	Jan Aug	Absolute	
°C	5	6	7	9	11	14	15	16	14	11	8	6	—	8 20	Jan Jul	
mm	96	76	51	56	61	54	79	76	76	86	107	110	**927**	2 12	−11 31	

Berlin, Germany. 52° 27' N. 57m														Max/min °C		
	J	F	M	A	M	J	J	A	S	O	N	D	Total	Jun Sep	Absolute	
°C	−1	0	4	8	13	16	18	17	14	9	4	1	—	2 23	Jun Sep	
mm	48	33	38	43	48	59	79	56	48	43	43	48	**609**	−3 13	−26 36	

for short periods, with continental influences increasing away from western coasts. This is evident even in such a narrow country as Britain, where there are also considerable rain-shadow effects (Figure 3.48).

There is precipitation through the year, the amount varying with exposure to westerlies and frontal rain belts and also with relief. Western parts of Europe mostly have a winter maximum, when depressions most frequently bring frontal rainfall. Occasionally continental high pressure slows the eastward movement of depressions, so that cloud and periods of rain persist in the west. Snowfall is seldom long-lasting, but, again, the duration increases towards continental interiors. In summer cyclonic systems are less frequent, though moist air drawn towards continental low pressure areas brings thundery storms over heated lowland.

In some *Cb* regions anticyclones are seldom sufficiently

Table 3.13

Invercargill, South Island, New Zealand. 46° 26' S. 4m														Max/min °C		
	J	F	M	A	M	J	J	A	S	O	N	D	Total	Jul Jan	Absolute	
°C	14	14	13	11	8	6	5	7	9	11	12	13	—	9 19	Jul Jan	
mm	107	84	102	105	112	91	81	81	81	105	107	102	**1155**	1 9	−7 32	

Cabo Raper, Southern Chile. 46° 50' S. 40m														Max/min °C		
	J	F	M	A	M	J	J	A	S	O	N	D	Total	Jul Jan	Absolute	
°C	11	11	10	9	8	7	6	6	7	8	9	10	—	8 14	Jun Mar	
mm	198	147	181	196	191	201	242	191	142	178	170	178	**2212**	3 8	−2 22	

Figure 3.47 Here on the northern Californian coastline tall redwoods and deciduous trees have been cleared for pasture. In this Cbf region the coastline northward from here to southern Alaska have forests of Douglas fir that survived the last ice-age.

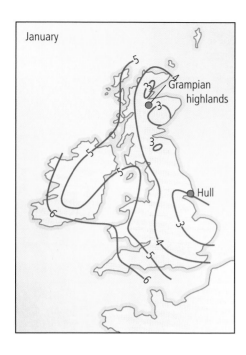

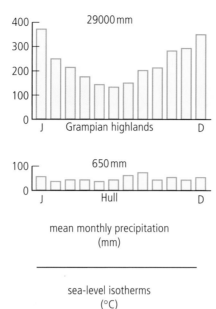

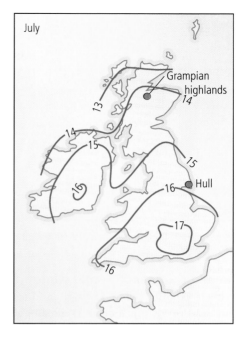

Figure 3.48 *Mean temperatures over the British Isles tend to decrease from east to west during winter, but from north to south during summer months, with the higher sun and longer days.*

established to control the weather for a lengthy period, especially in the southern hemisphere, where cyclonic systems and strong westerlies moving across the wide oceans meet narrow landmasses. Thus in New Zealand depressions rapidly cross the country, bringing short rainy periods followed by bright weather with much sunshine – though North Island occasionally receives storms of tropical origin. Southern Chile also experiences strong westerlies from a sequence of eastward-moving depressions.

In each of these regions the natural lowland vegetation is mixed deciduous–coniferous forest, with parent rocks favouring particular combinations with grasses. In Western Europe little *natural* forest remains, but in the western coastlands of the Americas there is still much coniferous forest with long-surviving species, such as the redwoods of northern California and the Douglas fir further north, and in Chile dense evergreen forests.

QUESTIONS

1 Explain how the direction of the winter and summer isotherms in Figure 3.48 are responses to (a) prevailing winds and frontal systems; (b) continental influences; (c) solar elevation; (d) length of daylight.
2 How does the precipitation regime (i) for the Grampian Highland station and (ii) that for Hull indicate effects, if any, of: (a) maritime influences; (b) topography; (c) a rain-shadow location; (d) continental influences?
3 Look at Figure 3.49 describe and account for (a) regional cloud cover; (b) north/south temperature differences; (c) the relative position of the warm front; (d) possible reasons for the Grampian precipitation regime shown in Figure 3.48.

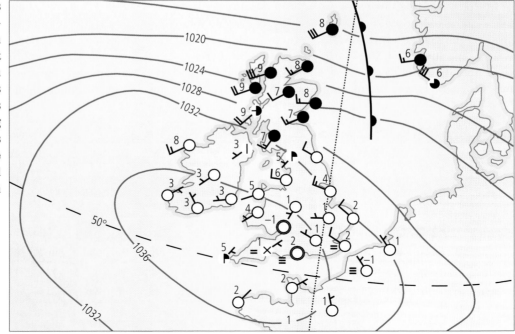

Figure 3.49 *Conditions during mid-winter, when for several days high pressure dominated weather over the British Isles, though with regional variations.*

Humid continental with cold winters
Da and Db

KEY

▨	Da warm summers	⣿	250 – 500mm p.a. precipitation·	remainder of shaded
☐	Db cool summers			area over 500mm p.a.

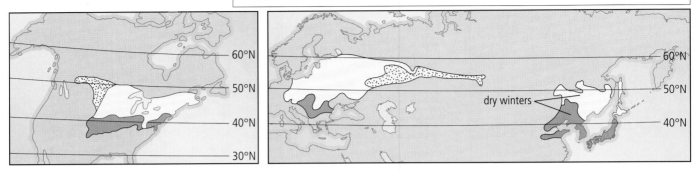

Figure 3.50 *These coastal and interior regions differ in precipitation received and in summer temperatures, but all experience incursions of cold cP air during winter.*

Humid continental – cold winter (*Da, Db*)

This is experienced in northern landmasses mainly between 40° and 60°N, but there are considerable variations due to latitude, location and distance from the ocean. The more southerly *Da* regions have warm summers. Those with a *Db* climate have cooler summers, but they also have longer periods with insolation. During winter, as indicated above, the western parts of Euro-Asia receive occasional outflowing cP air from the cold continental interior. There is rainfall or snow where it meets maritime air, but the amount of precipitation decreases eastward.

Over eastern Asia surges of dry, cold air from the interior are likely to raise dust, especially when their cold fronts cross the loess lands (Figure 3.7). In north-eastern China, and offshore, this cP air meets maritime polar (mP) air, creating occasional snowfalls. During summer the southern parts of this eastern Asian climatic region are very warm, with indrawn mT air bringing heavy showers.

In North America the winter weather in these regions is more changeable. Surges of cP air from the north give cold spells in the interior and snowfall where the cold air meets warmer maritime air over eastern and south-central areas. Here, too, the more southerly, and more maritime, *Da* climatic region has very warm summers and milder winters.

Figure 3.51 *In eastern Canada such hot summer days along this tributary to the St Lawrence give way to cold weather from October to April, with sub-zero temperatures during mid-winter.*

In fact in each of these regions summer temperatures are high for the latitude, which makes for convection storms in less stable inflowing air and gives them most of their precipitation at this time of year. Occasional depressions also bring rain to these parts of North America and even more frequently to the coastlands and islands of the Asiatic regions. In general the precipitation decreases poleward and towards the continental interior.

Table 3.14

Boston, Massachusetts. 42° 22′ N. 38m													
	J	F	M	A	M	J	J	A	S	O	N	D	Total
°C	−2	−2	2	8	14	19	22	21	17	12	6	−1	—
mm	71	84	96	89	79	81	84	71	81	84	71	86	**1036**

Max/min °C		
Jan Jul	Absolute	
2 27	Feb Jul	
−7 17	−28 40	

Churchill, Manitoba, Canada. 58° 47′ N. 13m													
	J	F	M	A	M	J	J	A	S	O	N	D	Total
°C	−28	−26	−21	−10	−1	6	12	11	5	3	−15	−24	—
mm	13	15	23	23	23	48	56	69	59	35	25	18	**406**

Max/min °C		
Jan Jul	Absolute	
−24 18	Jan Jul	
−33 6	−50 36	

Northern polar lands Dc and Dd: ET and EF

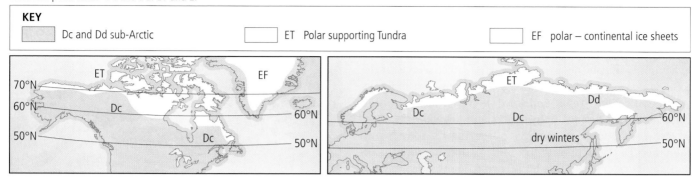

KEY

▨ Dc and Dd sub-Arctic	☐ ET Polar supporting Tundra	☐ EF polar – continental ice sheets

Figure 3.52 The climatic–vegetational divide between these sub-Arctic and Polar lands is where the northernmost territory with tree growth gives way to tundra.

Figure 3.53 Here amid Greenland's tundra, bordering its central icesheets, a small settlement amid grassy pasture is sheltered by a scree-fringed ridge.

Sub-Arctic (*Dc, Dd*)

The more northerly parts, especially, have long, extremely cold winters and short, but warm, summers. Acknowledging the vegetation response to climatic conditions, the poleward limit is the 10°C isotherm for the warmest month, below which tree growth is inhibited and tundra the main vegetation. Extensive softwood coniferous forest (taiga) is typical of the *Dc* region.

With continuous energy emissions through **winter**, the coldest monthly mean in the *Dc* region is below −38°C, with days below −50°C. Cold dense air masses accumulate. Atmospheric pressure is high, so at times air of low absolute humidity surges to lower latitudes and towards the coast.

Central Siberia becomes excessively cold in winter. Verkhoyansk in north-eastern Siberia's *Dd* region has recorded −68°C. But this is a broad climatic zone. In the south mean January temperatures, though low, resemble the −28°C of Churchill in Manitoba (Table 3.14).

During the short **summer**, with many daylight hours, the temperature climbs rapidly and at midday may approach 30°C. Mean temperatures vary with location, but at least one summer month has a mean above 10°C. Most precipitation comes in summer, though much of the light winter snowfall remains until the spring melt. Annual amounts are very small in and about the *Dd* region.

Table 3.15

Gällivare, Sweden. 67° 08′ N. 365m														Max/min °C		
	J	**F**	**M**	**A**	**M**	**J**	**J**	**A**	**S**	**O**	**N**	**D**	**Total**	**Jan Jul**	**Absolute**	
°C	−11	−12	−8	−2	6	11	15	12	6	−1	−1	−10	—	−7 21	Jan Jul	
mm	43	28	25	33	38	59	76	74	54	59	45	38	**569**	−16 9	−42 34	

Verkhoyansk, Russia. 67° 34′ N. 133° 51′ E. 100m														Max/min °C		
	J	**F**	**M**	**A**	**M**	**J**	**J**	**A**	**S**	**O**	**N**	**D**	**Total**	**Jan Jul**	**Absolute**	
°C	−51	−45	−32	−15	0	12	14	9	2	−13	−38	−47	—	−48 19	Feb Jul	
mm	5	5	3	5	8	23	28	25	13	8	8	5	**135**	−53 8	−68 37	

HIGH LATITUDE CLIMATES

Polar supporting Tundra (*ET*)

Here, where no month has a mean temperature above 10°C, is a tundra mat of herbs and dwarf shrubs, with stunted trees in warmer parts. The nature and abundance of plants varies with micro-relief and local micro-climates, so that rock faces may support only mosses and lichens.

Mean temperatures are above freezing for only two to four months. In winter, with no insolation for months on end, temperatures become bitterly cold. Snow remains until May, when melt-water held above permafrost produces swampy conditions. During summer, with the sun above the horizon for most of the 24 hours, days can be very warm and the diurnal range small. In continental interiors precipitation is mostly in summer, but seldom exceeds 250 mm a year. However, the near-coastal tundra-bearing regions in north-west Europe and the Labrador peninsula have rather more precipitation.

Polar icesheets and oceans (*EF*)

During summer, with long periods of daylight, the Arctic and Antarctic receive much solar energy, even though the rays are oblique and snow and ice surfaces reflect about four-fifths of that received. Their long-wave energy emissions are countered to some extent by summer in-flows of warm moist air, which release large amounts of latent heat as water vapour condenses to droplets and the drops solidify. By contrast during winter, the continuous energy loss makes for extremely low temperatures.

Over **Antarctica** surface temperatures vary with latitude and altitude. The mean annual temperature at Vostok, 3300 m above sea-level, is −56°C, compared with −51°C at the Pole, a thousand kilometres further south, but 600 m lower. Vostok's minimum recorded temperature is −88°C.

The estimated average precipitation, mostly small, hard snow particles, is the equivalent of 150–250 mm, though only about 50 mm near the South Pole. In **winter** there are few incursions of warm air from the strong encircling westerlies. Occasional eastward-travelling depressions bring heavy

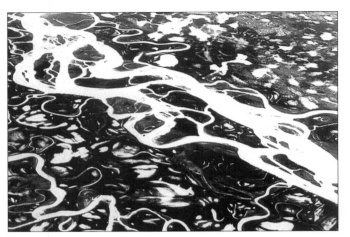

Figure 3.54 *In Arctic Canada rivers meander northward to the Beaufort Sea amid tundra and ice-scooped hollows. Its lakes, like the streams, are frozen during winter.*

snowfall to the continental margin, but few affect the central region. Inversion occurs in air above the cold surfaces, where strong down-slope winds develop, a hundred metres or so in depth, and with increasing speed create local blizzards.

In general, surface air moves outward towards the oceans. However, the high upper westerlies encircle a low pressure vortex (Figure 4.47), into which upper air currents carry natural and artificial pollutants, whose retention in the lower stratosphere and subsequent chemical actions are considered on p. 95.

From early **summer** there are inputs of solar energy. The earth is nearest the sun during the Antarctic summer, when its atmosphere receives some 7% more solar energy than the Arctic atmosphere in the northern summer. Inflowing moist air and occasional cyclonic systems may now move into the continent, which receives latent heat energy, and snow accumulates over the Antarctic icesheets.

The **Arctic**, apart from Greenland's elevated icesheets and narrow coastlands, is mainly an open ocean with a land perimeter, so that the climate is more variable. Surface short-wave reflection and emissions of long-wave energy are countered by energy inflows from warm air and by the reception of latent heat. These, with the modifying influences

Table 3.16

Little America, Ross Sea, Antarctica. 78° 34′ S. 163° 56′ W. 9m														Max/min °C	
	J	F	M	A	M	J	J	A	S	O	N	D	Total	Sep Jan	Absolute
°C	−7	−16	−21	−29	−31	−27	−38	−36	−40	−26	−19	−7	—	−34 −4	Sep Dec
														−46 −9	−59 6

Eismitte, Central Greenland. 70° 53′ N. 40° 42′ W. 3000m														Max/min °C	
	J	F	M	A	M	J	J	A	S	O	N	D	Total	Feb Jul	Absolute
°C	−36	−47	−40	−31	−21	−17	−12	−18	−22	−36	−43	−38	—	−32 −34	Mar Jul
mm	15	5	8	5	3	2	3	10	8	13	13	25	**110**	13 18	5 40

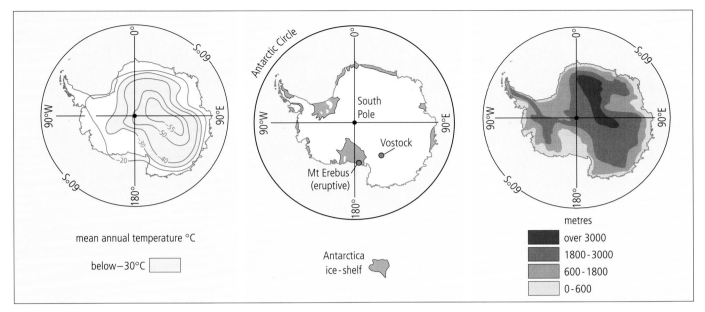

Figure 3.55

of oceanic surfaces of ice and open waters, keep air temperatures above those of Antarctica.

Mean annual temperatures on the Greenland icesheets are some −30°C to −35°C, compared with −20°C to −25°C over the surface of the Arctic Ocean. Atmospheric conditions in and around the Arctic are variable. At times cold cP air surges outward, gaining energy and humidity as mP air – both becoming involved in cyclonic disturbances, with frontal precipitation.

Taking both hemispheres together, ice makes up a significant proportion of Earth's surface, with a great contribution to planetary albedo as it reflects solar energy. As the ice surface increases, this should tend to cause global temperatures to decrease, and conversely a reduction in the ice coverage should allow more solar energy to be absorbed with a temperature increase. There is therefore some speculation about the tendency of global warming to effect an increase in the high Antarctic icesheet, which should have a counter effect by at least maintaining a reflective surface – though elsewhere a decrease in low-level ice coverage would act to encourage warming. These possibilities of ice increase or decrease are considered further in Chapter 4.

Figure 3.56 *Reflection of solar energy from extensive snow-covered landscapes, as here in Arctic North America and across Siberia, adds to that from the huge Antarctic and Arctic icesheets.*

Figure 3.57 This semi-arid country in the lower Indus valley, reclaimed for farming, became waterlogged as ground-water from a much irrigated area upstream moved down-valley beneath the surface. As local soils became saturated, saline water continuously rose to the surface, with salts acquired from rocks and soils as it seeped downstream. Evaporation caused it to become ever more saline.

BEYOND CLIMATIC CLASSIFICATIONS – LOCAL VARIATIONS

As indicated, it is often difficult to define regional boundaries in a classification based on what is required to support particular plants. The vegetation within the climatic region described may not be responding simply to the mean seasonal rainfall and temperature. It will often have adapted to conditions caused by *people's* reaction to the climate. Thus people may strive to overcome the effects of seasonal drought, but in doing so modify both the local climate and the plant life. Figures 3.57 and 3.58 show the effects of acting to combat the arid climate of the lower Indus valley in Pakistan by extensive riverside irrigation. This not only introduced new vegetation (crops), but created a downstream seepage of ground-water with acquired salts that made other land unsuitable for farming. There the vegetation is now mainly of halophytic shrubs on soils that are salty though moist – a change of vegetation and of near-surface climate.

Figure 3.58 Continuous evaporation left salt over much of the formerly cultivated surface. Salt-tolerant, halophytic shrubs established themselves in the saline, but moistened, soil.

Figure 3.59 *Below the snowy mountains of Kilimanjaro and Mawenzi well-grassed savanna on fertile soils among volcanic outflows supports numerous Maasai families. The seasonally migrating herdsmen move with their cattle, but about the semi-permanent settlements wide areas are now bare of grass, with trees removed or lopped for firewood.*

Figure 3.59 points to the problems of climatic classification and shows that natural vegetation may be drastically modified. In this broad climatic region, conditions on the snow-capped Kilimanjaro–Mawenzi heights vary with altitude. Close by the heights the usually abundant seasonal rain produces tree savanna, providing good grazing for Maasai herdsmen; however, large areas about their settlements are almost bare of vegetation.

Figure 3.60 shows where hillsides near La Paz and the landscape of crumbling pillars were once protected by woodland, which modified the climate. Today turbulence from exposed surfaces intensifies occasional storms, increasing both the run-off and the erosive impact of rain on the pillars. Together these examples show that within the broad climatic regions there are infinite 'patterns of micro-climates'.

Figure 3.60 *In high Bolivia summer storms followed by drought, with rapid evaporation at altitude, make clearance for cultivation hazardous – witness these gullied slopes near La Paz and the severely eroded foothills, with boulder–protected pillars of dry material, which in time collapse.*

SUMMARY

This chapter accounts for Earth's variety of climates, points to numerous variations within climatic regions and emphasises that most merge gradually into one another. It is possible to account for climatic differences between western Europe and Amazonia, but less easy to define the exact *Cb* boundaries in Europe, or even those of *Af* climate in western Amazonia, in view of the numerous variations bordering the eastern Andes.

The descriptions, maps and mean statistics give an overall view of conditions in designated climatic regions. Tables are included rather than graphs, for they reveal seasonal variations in more detail. However it is instructive to draw climographs based on the tables, for they visually emphasise contrasts between, say, *Af* and *Aw* climates. The tables and graphs relate, of course, to selected places, though statistics for other locations in the same climatic region may show considerable variations, as in Tables 3.3, 3.7 and 3.12. Remember that these statistics are not 'permanently' correct, but are ever-changing. Thus Chapters 4 and 5 examine why and on what scale, and debate reasons for global climatic changes.

QUESTIONS

1 Account for the differences in climate in maritime locations to the west and east of Euro-Asia at latitude 50°N.

2 Compare the west to east variations in climate in Equatorial latitudes in America with those in Africa, and account for the differences.
3 'Mountains make their own climates, and affect those of lands to the leeward.' Give examples related to the high ranges in the Americas.
4 Subsiding air affects the climate of (a) sub-tropical deserts and (b) mid-latitude interiors, but the cause and effects differ. Explain why subsidence is more permanent and aridity more established in *BWh* regions than in those with *BSk* and *BWk* climates.
5 Using vegetation as a response to climate makes it relatively easy to define the boundary between sub-Arctic and Polar climates. Give reasons for the difficulty of using this as a basis for climatic classification in lower latitudes.

PROJECT

Draw up a graphical model to show the soil-water budget for Los Angeles based on statistics in the table below:

1 Stipple the area showing withdrawal from water storage ($-G$) ... (between curves Ea and P).
2 Shade in blue the area showing the recharge of soil water ($+G$) ... (between curves P and Ep).
3 Shade in yellow the area showing the water shortage (D) ... (between Ep and Ea).
4 Give climatic reasons why there is a such a shortage and apparent necessity for irrigation (D); and why the symbol R, indicating a water surplus and possible surface run-off, is not included (see Figure 3.1).

BIBLIOGRAPHY AND RECOMMENDED READING

Asquith, N. and Whittaker, R., 1992, *Rainforest disturbance and recovery, Geography Review*, 5(5), 11
Atkinson, B., 1988, *Modelling weather and climate, Geography*, 83(2),147
Fairbrother, D., Tilling, S., Holmes, D. and Sanders, R., 1996, *Tropical rainforest in Malaysia, Geography Review*, 10(1), 20
Goudie, A., 1991, *The climatic sensitivity of desert margins, Geography*, 76(1),73
Henderson-Sellers, A. and Robinson, P., 1999, *Contemporary Climatology*, Longman
Money, D., 1996, *China in Change*, Hodder & Stoughton
Stott, T. and Stott, N., 1996, *Arctic Tundra, Geography Review*, 10(2), 18
Strahler, A. H. and Strahler, A. N., 1992, *Modern Physical Geography*, Wiley & Sons
Wheeler, D., 1996, *Spanish climate, Geography Review*, 10(1), 34

WEB SITES

Bureau of Meteorology Australia – http://www.bom.gov.au/climate
National Climatic Data Center (USA) – http://www.ncdc.noad.gov
National Oceanic and Atmospheric Administration (NOAA) – Climate Diagnostics Center (USA) – http://www.cdc.noaa.gov
South African Weather Bureau (South Africa) – http://www.sawbb.gov.za/

	J	F	M	A	M	J	J	A	S	O	N	D
P (mm)	79	76	71	25	10	3	–	–	5	15	29	66
Ea (mm)	+	+	+	42	36	25	19	14	17	18	22	+
Ep (mm)	31	38	49	60	74	90	118	116	91	68	52	37

P = precipitation; Ea = evapo-transpiration; (+ months where P exceeds Ea); Ep = potential evapo-transpiration.

Chapter 4
Why climates change – causes and effects

CYCLES OF CLIMATIC CHANGE

This chapter considers reasons for the variability of Earth's climates. We accept that weather is continually changing, though we recognise regular patterns – a month when fog is common, or a season when hurricanes occur. On the other hand, climate is usually regarded as 'permanent' and any change is viewed with concern. As we have seen, climatic statistics are presented as mean values, averaged over a period of at least 30 years – the mean annual temperature, the mean monthly rainfall for June. Variation from these values appears as an 'abnormal' change, often attributed to the effects of our increasing population using new technologies. For those who experience local effects of air pollution this appears a reasonable assumption. But climatic elements are continuously changing. Annual mean figures calculated a hundred years ago usually differ from today's statistics, though at the time they were seen as 'normal'. Considering what is happening on a global scale, it is easy to assume that an apparent change in one part of Earth's surface will be mirrored elsewhere. This may not be the case, though disturbance in one region can have repercussions across the globe (p. 89).

The scale of change – is there a norm?

Over hundreds, thousands and millions of years remarkable 'natural' climatic changes have taken place. There appear to have been large-scale cycles of climatic variation (Figure 4.4), and now, on a much smaller scale, fairly regular repetitive year-by-year variations of climatic patterns (Figure 4.12). This certainly suggests that it is unwise to regard our present climate as 'the norm', to be preserved at all cost.

It is also unwise to assume, as many do, that any climatic change will have disastrous effects environmentally and socially. Climatic change is not always detrimental to the landscape or to mankind. It may bring advantages in various ways to many locations – introducing more humid conditions to arid territory, for example.

It is essential to acknowledge, however, that we can, and do, affect climatic elements locally – some, possibly, on a global scale. So that the extent to which current changes are due to 'natural' or 'artificial' causes is debatable. Such issues are summarised in Chapter 5 and *should* be debated.

Landscapes reveal the ever-changing climate

Wherever one travels there is evidence in the landscape of long-term and short-term climatic changes. Figure 4.1 shows chalk downland, with gentle grassy slopes down to a hummocky valley. In fact only a short while ago, on a geological scale, this tranquil countryside was part of a peri-glacial landscape, with glacial outwash torrents above the permafrost layer eroding the main valley, and streams gulleying the slopes – evidence of a very different climate, with associated tundra-like vegetation. Subsequently Britain's natural vegetation changed as plants adapted to the warming climate.

Figure 4.1 Grazing maintains short grass over these slopes of the South Downs, in Sussex, where the gulleys indicate the action of outwash waters during the last glacial period.

Case study:
Evidence of climatic change

Figure 4.2 *Water cascades into Milford Sound at the head of this long fiord in the south-west of New Zealand's South Island.*

Figure 4.3 *A corner of Lake Eyre in 1976, when floods covered its normal vast expanse of salt.*

Warming over 12 000 years

In every part of the world there are indications of long-term and short-term climatic changes. The evidence of recovery from very cold conditions in the landscape shown in Figure 4.1 is also apparent at the head of Milford Sound in the south-west of New Zealand's South Island (Figure 4.2). Here the abundant rain and cascading run-off in this humid, mountainous coastal region are brought about by the continuous sequence of strong, mild westerlies and low pressure systems that pass freely and frequently across the wide southern oceans. But Milford Sound is a long fiord and at one time this maritime location in the 'Roaring Forties' made for heavy snowfalls. Here the low air temperature allowed ice to accumulate and feed the large glacier that smoothed these contours and eroded this now drowned valley. Since then global warming has led to conditions very different from those 12 000 years ago, and for a long while before that (Figure 4.9).

Short-term change – amid evidence of past conditions

Climatic change is not necessarily a gradual trend over a long period. In many parts of the world what appear to be typical features of the climate are abruptly disturbed by a period of unusual weather, before reverting to so-called 'normal' conditions – as with the El Niño events.

Figure 4.3 shows a corner of Lake Eyre in the interior of South Australia with its wide depression uncharacteristically flooded. For most of the time Lake Eyre is some 8 000 square kilometres of salt flat, 12 metres below sea-level. But occasionally, as here in 1976, the run-off from torrential rains pours through the old dry river systems of a wetter climatic period and fills this vast depression. This seems to occur irregularly at intervals of several decades, and hence as a climatic feature.

There are also indications of long-term climatic changes – both by the former stream systems, telling of more humid conditions, and by the long red parallel dunes, which can be seen around the lake. These also occur over wide areas of Australia's ancient interior tablelands, indicating both the aridity and different wind directions in the distant past.

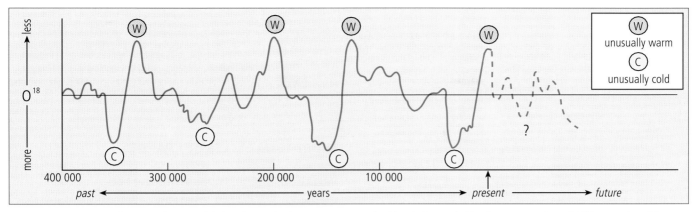

Figure 4.4 *Long-term climatic cycles on a global scale – with alternating warm and cold periods revealed by the concentration of O^{18} in oceanic residues.*

LONG-TERM CYCLES OF CHANGE

The behaviour of Earth itself

Figure 4.4 shows that during the past half million years there have been cycles of alternating warm and cold climates, revealed in this case by the proportion of a heavy isotope of oxygen O^{18} in the remains of minute foraminifera on the ocean floor. When these lived near the surface they obtained a higher proportion of O^{18} during colder conditions.

Assuming, incorrectly, that the sun's energy output remains constant, there are reasons why the energy received at Earth's surface varies over specific periods. Over the ages the distance between Earth and sun changes, as does the altitude of the noonday sun in all parts of the globe. Over a 110 000-year period **Earth's orbit about the sun** varies from more elliptical to more circular (Figure 4.5). This is significant even in the short term, because the present elliptical orbit makes for considerable contrasts in energy received in the Arctic and in Antarctica during their summer months, as Earth is closer to the sun in January than in July.

The energy received at any particular latitude also changes, for the angle between Earth's axis of rotation and the plane in which it moves about the sun varies by some 2·5° during a 40 000 year cycle. As the tilt increases, so do the seasonal contrasts. Also, as it spins, the Earth wobbles. Its axis swivels over a 10 500-year cycle, which also affects the angle of the noonday sun across the globe.

The combination of these spatial variations, shown in Figure 4.5, must affect both local weather and global climates. In fact over hundreds of millions of years the Earth's surface has experienced major climatic changes. About 60 million years ago there were no polar icesheets. Then global temperatures decreased over a long period. Some 14 million years ago ice began to cover Antarctica, though the northern

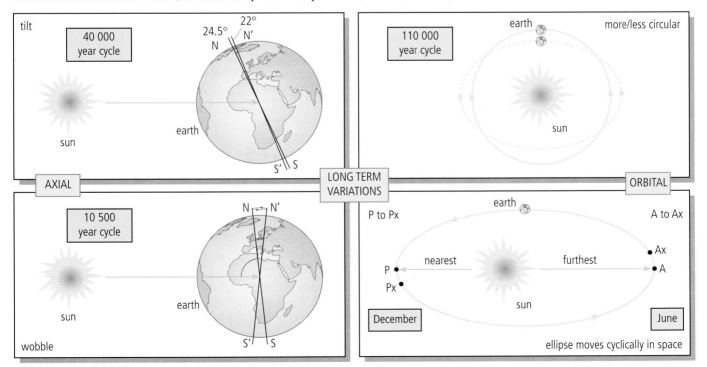

Figure 4.5 *Periodic variations in Earth's orbit about the sun and in the angle of its axis to the ecliptic plane – affecting the amount of energy received seasonally at any latitude.*

icesheets fully developed only two million years ago. During the last two million years cold glacial periods and warmer inter-glacial periods have alternated.

This cyclic pattern (Figure 4.4) suggests responses to the variations in Earth's orbit and to periodic changes related to Earth's axis of rotation. It is remarkable that for only a small fraction of the time has the global climate been as warm as it is today. It is also interesting to contemplate further changes suggested by the broken line, especially as reversion to glacial conditions can occur over very few years.

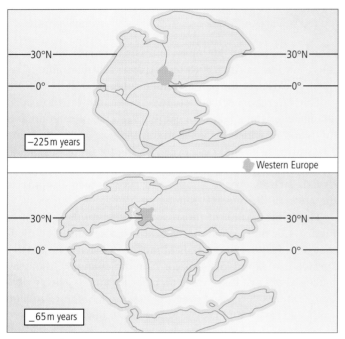

Figure 4.6 *Variations in regional climate over the ages may be due to 'continental drift', as Earth's crustal plates slowly move. Thus the climate of western Europe will have responded to different latitudinal influences in Permian times (225 million years ago), at the beginning of the Tertiary period (65 million years ago), and in its present location.*

Surface movement towards a different climatic region

There are other reasons why over time a land area may experience climatic change. For instance, **crustal plate movements** may cause an area to migrate slowly to other latitudes with different climatic characteristics. Collisions

between plates can also affect local and global climates. Thus some 40 million years ago the plate carrying India met the Asian landmass and slowly continued to create the Himalayan ranges and elevate the Tibetan plateau. In the process it disturbed atmospheric circulations over much of Euro-Asia, with wider effects on global climates.

SHORT-TERM CYCLES OF CHANGE

Variations in emissions from the sun

During Earth's existence the sun's luminosity has increased by a third. There are continuous fluctuations in solar emissions and in energy reaching Earth, now monitored by satellites. At times exceptional flares of energy are emitted from the sun's surface through faculae – bright areas associated with cooler, dark sunspots. A varying number of sunspots appear each year, groups of convective disturbances, often many thousands of kilometres across. Their incidence has been recorded over the centuries. Annual maximum and minimum occurrences each appear to take place in a cycle of about 11 years, with a trend for the maximum number to increase for about 80 years and then abruptly decrease. Figure 4.7 shows such an upward trend, followed by a decrease after the 1950s into the 1970s. In recent years intense flares have accompanied sunspot groups.

As the amount of solar energy entering Earth's atmosphere increases there should be climatic responses. Changes in mean air temperature do seem to have matched observed fluctuations in solar emissions, especially in the northern hemisphere. The processes that might cause them to respond to such influxes are not entirely understood. However, similarities between sunspot cycles and fluctuations in mean air temperature are striking. This is well illustrated in the correspondence between the remarkable decrease in sunspot activity between the sixteenth and late seventeenth centuries – the 'Maunder Minimum' (Figure 4.8) – and the unusually cold years of the 'Little Ice Age'. But, as yet, there is no clear evidence of the cause of such coincidence.

More recent climatic changes

Over the last ten thousand years human civilisation has developed during a recovery from glacial conditions (Figure

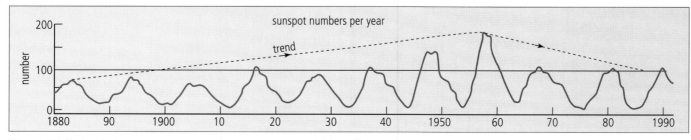

Figure 4.7 *During the last century sunspot numbers per year have varied from about 50 to 200, regularly waxing and waning, with maximum numbers at about 11-year intervals. A trend for maximum numbers to increase over the decades, followed by a sudden decrease, has been observed over the centuries.*

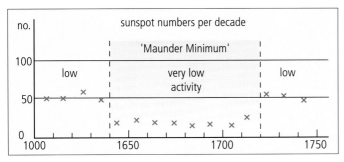

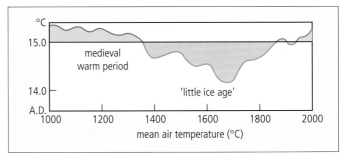

Figure 4.8 *Very few sunspots occurred during the sixteenth to eighteenth centuries, as shown by the numbers appearing during ten-year periods (compared with the annual numbers in Figure 4.7). Thus accompanying outbursts of energy were exceptionally low over the period of the 'Little Ice Age' (Figure 4.10).*

Figure 4.10 *From the warmth of medieval times the mean air temperatures of the northern hemisphere fell to those of an exceptionally cold period, followed by a recovery during the eighteenth to twentieth centuries.*

4.9). During this warmer period mean sea-level global air temperatures have fluctuated by little more than one degree Celsius, and over the last thousand years there have been fluctuations within this 1C° range (Figures 4.9 and 4.10).

As indicated, in Britain medieval warmth was followed by a remarkable decline in mean annual temperature over the period referred to as a 'Little Ice Age' – though even then the weather varied, with occasional warm years. Through the eighteenth and nineteenth centuries, and into the twentieth, mean temperatures have been recovering from that exceptionally cold period, though with fluctuations and some remarkably cold spells. To some extent the rise in global

mean temperature of about half a degree Celsius between 1880 and 1980 could be viewed in relation to this continuing recovery, especially in the northern hemisphere – for the southern oceanic hemisphere, though much cooler, was less affected during the Little Ice Age.

Figure 4.12 shows that as mean air temperature increased from 1880 to the 1950s, there were alternating cooler and warmer spells. From the 1950s mean temperatures declined, with some very cold spells during the early 1960s and into the 1970s, with many predicting further decline towards an Ice Age. Since then there have been fluctuations, with very warm spells in northern temperate regions in 1976 and during the end of the 1980s, and in the late 1990s there have been a sequence of particularly warm winters.

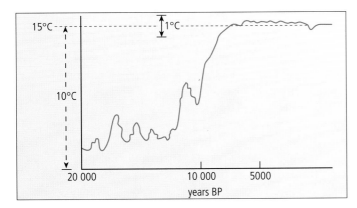

Figure 4.9 *The abrupt rise in mean temperature following a series of 'Ice Ages' contrasts with lesser fluctuations during the current warm period, which have been within a range of 1C°.*

Figure 4.11 *Icicles, rime and snow on Exmoor in 1963, during a period when northern hemisphere's mean air temperature sharply declined.*

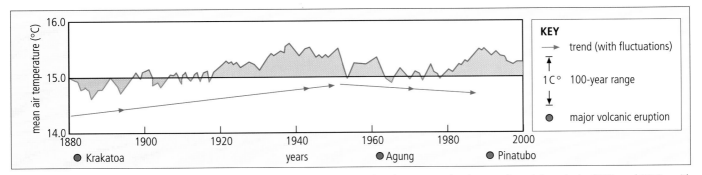

Figure 4.12 *Mean air temperatures during the past century show an upward trend to the 1950s and a downward trend through the 1960s and 1970s, with fluctuations since. During the 1990s northern winter temperatures tended to rise. The major volcanic eruptions shown appear to have been followed by exceptionally cool years.*

ATMOSPHERIC WARMING

The greenhouse effect

It is accepted that mean global temperature has increased by some 0.6C° over the past century, with indications that warming may continue for a number of decades. Possible consequences are considered on pp. 82 to 89 and are a subject for debate in Chapter 5, for human activities can stimulate the warming process through the greenhouse effect. Unfortunately the media often protray the 'greenhouse effect' as something unpleasant, even though it is **the essential process** that raises air to a temperature to which life-forms adapt. In the troposphere water vapour is the most efficient absorber of long-wave energy, taking up about 97% of that emitted from the surface. Of the rest some 3% is absorbed by carbon dioxide and relatively small amounts by other atmospheric gases. All re-radiate energy to other molecules about them, and hence are instrumental in so-called 'global warming'.

Figure 4.13 *One of the numerous volcanic solfatara in St Lucia emitting gases to the atmosphere, its rims stained with sulphur. Apart from occasional massive eruptions, thousands of volcanic outlets across the globe annually emit huge quantities of gases and particles. Major eruptions hit headlines, but vulcanicity is always a prime source of natural atmospheric pollutants.*

Natural emissions act to modify warming

Apart from varying solar inputs, other natural phenomena appear to have considerable effects on air temperature. Among these are the continuous gaseous and particle emissions into the atmosphere from thousands of active volcanic sources. Occasional eruptions on the scale of Krakatoa, Agung and Pinatubo considerably increase the volumes emitted. Volcanic particles both absorb and scatter energy and act as nuclei for cloud formation, with consequent cooling effects. After a major eruption huge emissions of sulphur dioxide form small sulphuric acid aerosols high in the troposphere, which absorb and scatter solar energy.

Their passage across the globe may affect regional temperatures for a number of years.

Periods of exceptional world-wide volcanic activity, not simply huge eruptions, have been followed by cold, wet years, as in the northern hemisphere in the 1760s and 1840s. The opposite has also been noted – during the warming period from the 1920s to 1940s there was a notable decrease in volcanic activity. Atmospheric dust from arid regions also varies in quantity and may act to lower air temperatures, both directly and as cloud-forming nuclei.

Carbon dioxide – its role in global warming?

If curbing global warming is seen to be an international priority, with serious reduction in carbon dioxide emissions from artificial sources aimed at lowering its atmospheric concentration, it is essential to consider carefully the various roles the gas plays in the atmosphere.

Inset 4.1
Carbon Dioxide

- Carbon dioxide (CO_2) is a non-poisonous gas that absorbs long-wave energy only from wavelengths 4.5µm and 12–18µm. Peak absorption is at 14.7µm – a wavelength of strong energy emissions from the surface.
- It is soluble in water. At 15°C water can dissolve about its own volume of the gas, and its ability to retain carbon dioxide increases as the temperature falls.
- Absorbed by plants containing chlorophyll, light energy enables it to react to form carbohydrates. Its carbon is thus stored in the world's biomass and in the products of plant decay (such as coal).
- It is absorbed by oceans, where minute plants (phytoplankton) transfer it to marine life-forms, whose shells and organic debris sink to accumulate as deep carbonate sedimentaries and hydrocarbons (fossil fuels). However, the rate of intake is slow relative to the rate of recent additions to the atmosphere.
- With water it forms dilute carbonic acid (H_2CO_3), which chemically weathers rock, especially limestone.
- It enters the atmosphere as plants and animals absorb oxygen and exhale carbon dioxide and from decaying organic matter and weathering limestone.
- It is now released in increasing volumes by the combustion of fossil fuels.
- If the atmospheric carbon content is taken as 1.0, carbon is present in the following proportions: **on land** (in plants 1.2; in humans 1.8); **in oceans** (in surface water 0.9; in deepwater organics 2.4; in deep sedimentaries 54.0).

Carbon dioxide is, of course, emitted from natural sources, such as respiration, plant and animal decay, and limestone weathering. However, contributions from domestic and industrial fuels, manufacturing, transport, and farming practices are increasing. Reduction in vegetation affects CO_2 exchanges. Deforestation may transfer much of the carbon from trees to the atmosphere; replanting allows it to be absorbed over the growing period.

As carbon dioxide warms the air about it, additional amounts should increase atmospheric temperature. However, it only absorbs long-wave energy emissions of certain wavelengths and in relatively narrow bands. In the opinion of some scientists the present concentration of CO_2, in conjunction with water vapour, already absorbs most of the energy emitted in those wavelengths, suggesting that additional inputs of the gas might have relatively little warming effect. Nevertheless, of atmospheric gases, it is the main absorber of long-wave energy, though far second to water vapour.

In the distant past there have been close relationships between exceptional global temperatures and the atmosphere's CO_2 content. About 100 000 years ago, with air temperatures slightly higher than at present, atmospheric CO_2 concentration was some 300 ppm. However, during the last Ice Age, some 20 000 years ago, it was only 190 ppm.

Even so, such apparent correlations should be examined further. For instance, the colder the water the more readily CO_2 is retained, so that during a very cold period the oceans, which cover 70% of Earth's surface, would hold more and slowly transfer it to the depths. During a hotter period warmer oceans release CO_2 more readily, and a higher atmospheric CO_2 concentration would tend to boost air temperature through the greenhouse effect. There are also indications that for some of these exceptionally warm periods the temperature increase *preceded* the rise in CO_2 concentration. Whatever the effects of carbon dioxide, it is unlikely to have been the sole cause of such excessive global warming/cooling.

Controlling emissions of 'greenhouse gases'

Natural sources and human activities release other gases that promote atmospheric warming. However, with low concentration, each contributes relatively little to the greenhouse effect. Among them is methane, whose emissions from cattle herds, waste tips and production of padi rice add to that released by natural wetlands, termites and ocean sources. It seems sensible to control its emissions, along with others such as the nitrogen oxides, chlorofluorocarbons (CFCs) and ozone – in their case, perhaps, as much to curb obnoxious pollution and chemical effects as their contribution to global warming.

The appeal 'reduce carbon dioxide' is so widely made as to become a catch-phrase, though the extent of its contribution to global warming over the past century is still debatable. Controlling its concentration seems prudent, though the conception of it as an environmentally disruptive 'villain' needs re-appraisal. Apart from its essential exchanges with life-forms, research shows that additional atmospheric carbon dioxide tends to increase plant productivity though photosynthesis, and by acting to close leaf stomata it reduces water loss through transpiration.

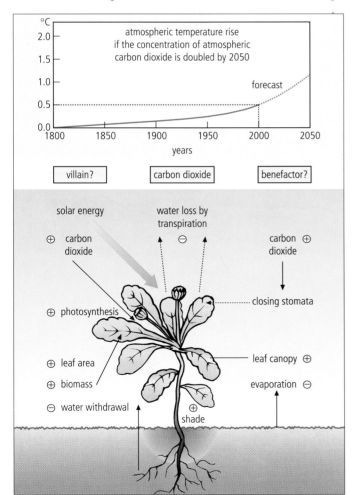

Figure 4.14 Above: *the rise in mean annual temperature since 1800, with a computer-generated view of an increasing rate should CO_2 concentration be doubled over fifty years.* **Below**: *the effects on plants of increasing the concentration of CO_2. This was found, experimentally, to increase the biomass and reduce water withdrawal from soils, to the benefit of vegetation and crops.*

QUESTIONS

The extent to which global warming is due to natural processes or to our own inputs is debatable.

1 Point to causes of climatic change that occurred when human inputs could not have had any noticeable effect.

2 Consider the scale of atmospheric warming/cooling over the past century. Point to periodic fluctuations that suggest human inputs are unlikely to have been the sole cause of an overall warming.

3 Suggest why during the Little Ice Age conditions in the southern hemisphere were much less severe than in the northern. Give reasons why an observed temperature increase in one region may not necessarily be mirrored elsewhere in the world.

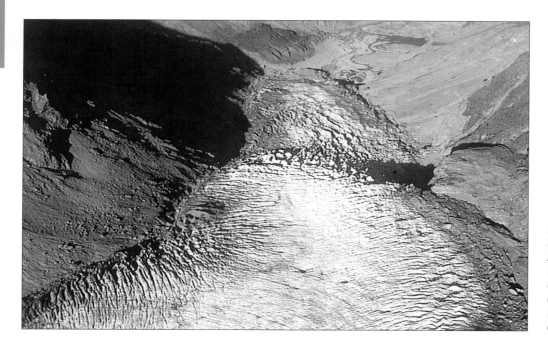

Figure 4.15 *Recent warming has caused low-level glaciers in New Zealand's Southern Alps to retreat far up their U-valleys. Morainic deposits like those about these outwash streams at the snout of the Fox glacier provide evidence of their former extension.*

COMPLEX EFFECTS OF GLOBAL WARMING

Effects on glacial accumulations

The rise of 0.5–0.6C° over the past century has been accompanied by a retreat of glaciers in many parts of the world and by the melting of fringe ice, with the detachment of ice shelves about Antarctica. This has not been a uniform effect – glaciers in Scandinavia and parts of the Himalaya have recently been advancing.

Research, by the Swiss in particular, has shown that over the centuries warm periods with high inputs of solar energy *have* caused low level ice to retreat, whereas at high levels the heavier snowfall has accumulated and acted to increase the mass of the upper icefields. Conversely, during cold periods with minimum solar activity the mass of low-level ice has increased, while that of high icefields has tended to decrease (Figure 4.16).

It is widely asserted that global warming is likely to remove polar icesheets, accompanied by a huge rise in sea-level. Over the last half-century, however, it appears that the mass of many high ice sources has been gradually increasing, including the Antarctic icesheets, with their mean thickness of 2.5 km over an area the size of the USA. This view is supported by recent surface sampling and satellite observations. In winter air sinks over Antarctica and there is relatively little snow, but falls increase with the advent of summer. A warming ocean, providing additional moisture, makes for a heavier snowfall, likely to bring a gradual increase in the mass of the icesheets (Figure 4.17). At low levels, however, more heat energy causes peripheral ice to melt more rapidly and iceshelves to break away more frequently. As detached floating ice melts, it does not cause a sea-level rise, but the

slip of a land-based ice mass into the ocean would slightly add to the volume of water, on a global scale.

In the Arctic, Greenland's icesheets and peripheral ice are more vulnerable to air temperature increase, and surface winds carry energy more freely over the Arctic's expanses of ocean. Retreat of low-level glaciers releases cold water,

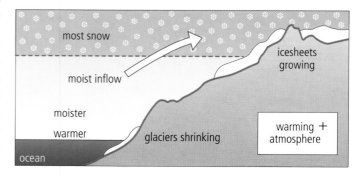

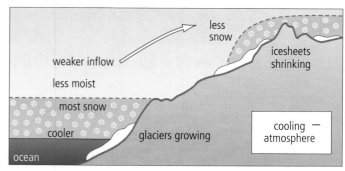

Figure 4.16 *Above*: *During warm periods, while low-level ice melts and glaciers retreat, at high level inflows of moist air give heavier snowfall and tend to increase the volume of icefields, as observed in both Switzerland and Alaska.* *Below*: *During cold periods heavy snow at low levels, and the lower rate of melting, causes glaciers to advance, whereas the volume of high snowfields, receiving less snow, tends to decrease.*

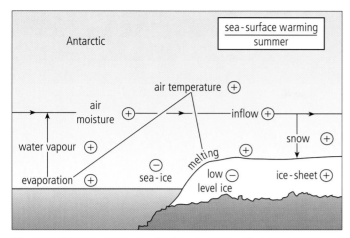

Figure 4.17 *Atmospheric warming increases sea-surface temperatures, releasing more water vapour and making for heavier snowfall from early summer, adding to the mass of the Antarctic icesheets. However, low-level ice melts more rapidly – hence concern about the West Antarctic iceshelf (Figure 3.55) which extends offshore, where it is grounded below sea-level.*

reducing the salinity and rate of sinking of adjacent North Atlantic waters. This may affect their deep-water circulation, which influences the flow and location of the North Atlantic Drift and thus brings colder conditions to western Europe.

Causes and effects of sea-level rise

With global warming both the expansion of ocean water and the melt-water from retreating glaciers make for a rise in sea-level. There has been a 10–20 cm rise over the last century and therefore concern over a future increase in coastal flooding, though such a slow rise allows time for protective measures. The rate of rise will be crucial. Predictions have been wildly exaggerated in the past – in the 1980s environmental groups, and textbooks, were suggesting rises of 10 metres per half-century! The present rate may indeed

increase, but when considering potential sea-level rise and its effects, other factors must be appreciated:

- Ocean volume can be affected by crustal movements; and even a small increase in ocean area due to plate movements could lower sea-level considerably.
- The possibility of coastal drowning may depend on the nature of the coastland, tectonic uplift or lowering, or movements due to recovery from depression by icesheets.
- Near-coastal areas may subside because of continuous extraction of water or oil; as in greater Los Angeles, where considerable annual surface subsidence continues (Figure 4.18).
- A slow sea-level rise is of little danger to low-lying atolls as their outer reefs are maintained by upward coral growth. However, a very fast rate might cause concern.
- Any increase in the frequency and intensity of typhoons in conjunction with a small sea-level rise might endanger low-lying coastal areas in the outer tropics.

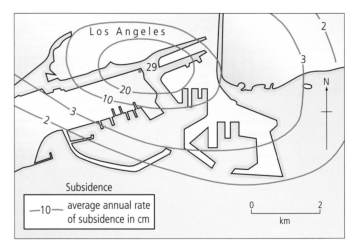

Figure 4.18 *The annual rate of subsidence affecting Los Angeles for many decades.*

Figure 4.19 *Los Angeles, where sea-walls protect the harbours. Subsidence is a problem for the high-rise centre, across this inlet.*

Figure 4.20 In Bangladesh, south of the floodplains of the river estuaries, flat deltaic islands extend into the Bay of Bengal, their alluvial soils inviting cultivation by families from the over-populated and seasonally inundated floodplains inland.

Flooding in Bangladesh

It is often asserted that sea-level rise may mean disastrous flooding of coastal lowland world-wide, but the situation in each region needs careful examination, as in the case of Bangladesh. From time to time cyclone-induced surges in the Bay of Bengal flood the delta lands, with disastrous consequences for the increasing number of settlers from the mainland. It is widely assumed that such flooding is an increasing hazard for Bangladesh as a whole, and that rising sea-level is therefore a major threat. Yet the United Nations Disaster Project Report of 1990 stated that 'there is no evidence for a "greenhouse" induced rise in sea-level posing a threat'. Any sea-level rise might theoretically affect the height of storm surges, but the scale is such that any significant effect is unlikely.

The delta is still extending, which sadly gives an incentive for even more of the growing population to clear, settle and farm – only to suffer during the next major cyclone-induced flooding. The inevitable publicity attending such periodic disasters overshadows the fact that during each wet monsoon more than a fifth of the country is inundated by overflows from rivers draining huge catchments, again with loss of life and disastrous effects on crops and settlements. The current project-planning for flood control and damage limitation (FCD) concentrates mainly on river-flood protection, but with coastal polders protected from storm surges (Figure 4.21).

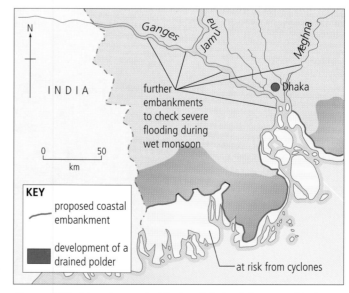

Figure 4.21 A long-term project for Bangladesh, aiming to provide river embankments, pumping facilities, storages to control the annual flooding, and coastal embankments to protect reclaimed farmlands from exceptional cyclone-induced surges.

QUESTIONS

Point out the problems of predicting the extent of global warming by:

a using a regional retreat of ice as evidence.

b the periodic flooding of a coastland area.

c observing unusual seasonal temperatures over a particular area.

Figure 4.22 Eritrea in the early 1970s when years of severe drought left streams dry, and cultivation about such settlements became impossible.

Effects on arid marginal lands

There is still doubt about the ability of computer-assisted climatic models to predict with any certainty the extent of global warming or likely effects in specific parts of the world. They have generally suggested that global warming may lead to drier conditions over central Asia and western-central USA. Also they have indicated that, as much of the northern hemisphere warms, there would be increasing climatic instability, especially in maritime regions.

Industrially developed countries should be able to deal with slow climatic change by using advanced technology – introducing suitable varieties of crops, for instance – but the developing countries would find it more difficult to deal with adverse conditions.

In areas such as the Sahel, bordering the Sahara, any trend towards drier conditions should exaggerate the problems periodic droughts pose for its semi-nomadic, and occasionally migratory, populations and for settlers in the adjoining dry savanna. However, for the Sahel and other semi-arid lands, including those in south-west Africa and Western Australia, there have been predictions of moister rather than drier conditions, though with some counter-effects. While increasing convection should encourage the inflow of moist air from over the oceans, higher temperatures and more rapid evaporation might dry out surfaces, and heavier storms erode them. Thus predictions of global warming and possible consequences need very careful examination. Beware of the danger of using buzzwords, such as 'desertification', related to these marginal lands, for they mask what are complex processes.

Figure 4.23 Eritrea during 1975 when drought was temporarily broken and, as here, moist air moving in from the Red Sea greened the hillsides. The rate of recovery of these marginal lands and the resilience of the population are striking.

Inset 4.2
Variability

In the semi-arid lands the amount of annual rainfall varies considerably, as does rainfall intensity.

- Rainfall variability (%) is measured by dividing the mean deviation from the average by the average and multiplying by 100. The mean average variability for Alice Springs in the Australian interior is 55%, compared with London's 10%.

- Very high rainfall can be recorded in short periods and in individual storms. In 1969 El Djem in Tunisia received 319 mm in three days, its mean annual rainfall being 275 mm.

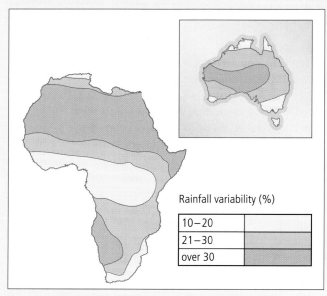

Rainfall variability (%)	
10 – 20	
21 – 30	
over 30	

Figure 4.24 Rainfall variability over Africa and Australia.

When lengthy drought affects these lands, the term '**desertification**' is frequently used to suggest a permanent extension of desert over its margins. It is also used to imply that the overall extent of the world's deserts is increasing. In reality when a moister period does replace prolonged drought, most desert margins recover surprisingly quickly (Figures 4.22 and 4.23). Nevertheless during such drought people tend to move to adjacent habitable territory, where even a temporary population increase can cause considerable land **degradation** – as in the Sahel.

Unpredictable effects of global warming

The main difficulty in predicting the effects of global warming is that places with allegedly similar climate respond differently because of locational or physical variations. While the West African desert borderlands may benefit

Figure 4.25 Farming the floodplain and land about this tributary to the upper Tigris is precarious under the present variable climatic conditions. The extent to which continued atmospheric warming would affect the balance between available water and the rate of evaporation is extremely difficult to assess, for reasons given below.

as moist air is more readily drawn in from the oceans, this may not be the case in the Middle East.

Figure 4.25 shows cultivation by small rural communities concentrated on the floodplains of tributaries to the upper Tigris. Assessing likely effects of atmospheric warming on these precarious settlements means considering possibilities that do not apply to the African Sahel.

Direct maritime influences are less, but during winter the cyclonic systems from the west may become more frequent and more active, bringing more rain.

However, people depend on small irrigation networks, so that higher temperature and increasing evaporation could make farming floodplain margins more precarious.

The rivers receive much water from snow-melt in the northern mountains. Warming may well cause the snow masses to decrease – and, unlike the high Alaskan snow-fields, they may not receive extra inflows of moist air – so that supplies for the rivers may become inadequate.

Similarly other parts of the world sharing a particular type of climate are likely to respond to 'global warming' in different ways.

QUESTIONS

1 Suggest why the upper Tigris and Euphrates rivers and their tributaries have maximum flow during the winter period, but then a sudden large discharge during spring and early summer.

2 How does this relate to the observation that 'the larger the proportion of highland in the catchment of these rivers and streams the more pronounced the spring and summer peak flow'?

Case study: The Sahel

Throughout the Sahel fluctuations between periods with adequate rainfall and prolonged periods of drought have been occurring over the centuries. During years with adequate summer rain herdsmen and their families establish homes on the Saharan fringe as bases for seasonal transhumance. When air subsidence persistently prevents summer incursions of moist air, drought occurs (Figure 4.28), forcing the families, with their livestock, to migrate southward. There, during prolonged drought, they add to the already growing population of these now very dry savannas, stripping bushes and trees for firewood and, with the extra concentration of cattle, creating surface erosion about the wells.

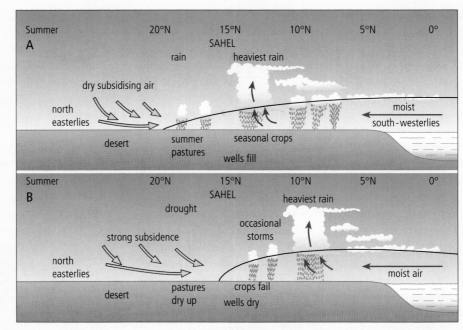

Figure 4.28 *At times over successive years moist air indrawn over West Africa fails to reach the marginal desert lands of the Sahel (B), which continue to receive hot, dry air from the sub-tropical high. During such a period there is widespread southward migration.*

At such times there is undoubtedly much degradation within these lands, yet little of the arid margins is permanently converted to desert. The speed with which plants re-establish themselves as rains return is remarkable – after a single wet summer there may be sufficient vegetation over the desert fringe for sporadic grazing, and after a year or two with sufficient rain there may be semi-nomadic pastoralism once again.

In the adjoining dry savanna lands, where some areas remain despoiled by such temporary migration, the continuing large natural increase in the permanent population is probably a more immediate environmental threat than the long-term possibility of global warming extending periods of drought.

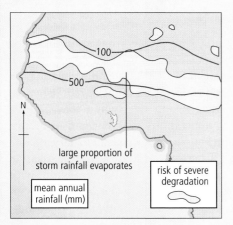

Figure 4.26 *The isohyets show amounts capable of supporting transhumance along the desert fringe, though, as Figure 4.27 shows, 'mean' figures have their limitation.*

QUESTIONS

1 Explain the difference between the terms 'degradation' and 'desertification'.

2 How are circulations within the Hadley cell (Figure 2.10) linked to movements of the dry northerlies over the Sahel?

3 Consider why threats to the natural environment are often closely linked to the rate of population growth, and how this relates to conditions within the Sahel. Examine similar relationships in pressures on the *Af* rainforests in parts of Indonesia and Brazil.

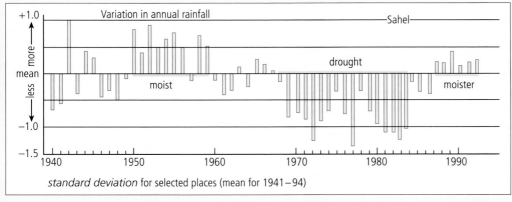

Figure 4.27 *In the Sahel periods with erratic rainfall and lengthy drought have alternated over the centuries – behaviour emphasised by this plot of standard deviation from the mean rainfall between 1941 and 1994.*

Warming effects on permafrost

Permafrost – perennial frozen ground, often to a depth of 300 m to 1100 m below tundra and Arctic forest – is mostly an inheritance from the last Ice Age, though some has developed under the existing climate. In summer there is shallow surface thawing, at the most a metre deep. The effects of settlements, highways and oil pipelines have shown the fragility of surfaces normally protected by a moss or peat layer. The removal of this layer makes for deeper summer thaw and allows sub-surface ice to melt, leaving an uneven surface with muddy channels.

Townships adapt to permafrost by erecting buildings on deep piles, with insulating air space below. Their facilities like water, sewage and heat from a communal source are delivered through pipes encased in an above-ground utilidor (Figure 4.29). Urban warmth in some towns has caused buildings to sink, and further atmospheric warming could be a problem. On the open tundra it would allow shallow winter-frozen lakes to cover wider areas. However, as already evident, trees would begin to colonise the southern margins. As the frozen soils contain abundant carbon compounds, permafrost melting could release carbon dioxide and methane, increasing the atmosphere's 'greenhouse gas' content. All of this depends on the extent to which any

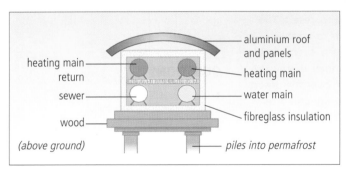

Figure 4.29 *Many settlements in permafrost areas rely on* above-surface *utilidors like this to provide vital serivces.*

overall global warming would affect this continental climatic regime of high summer temperatures for the latitude and very low winter ones.

Of course, natural warming at the end of the last Ice Age removed permafrost from large areas of Europe. Its southern limits are roughly that of the −5°C mean annual isotherm, so that permafrost conditions were experienced over most of mainland Britain. In fact in East Anglia crops marks, or air photos, show polygonal patterns left by ice wedges that tapered deep into the ground. Enlarged, as water seeped down in summer and then refroze, the widening cracks formed polygons. Firm evidence of climatic change.

A periodic climatic disturbance – El Niño

A severe El Niño event is apt to be attributed to global warming, but its causes and effects are complex. This is a periodic disturbance whose intensity has varied over the centuries, though always likely to bring disastrous floods to coastal Peru, with associated droughts in other tropical and sub-tropical locations.

Over the Pacific to the south of the Equator there is a large-scale climatic system – the Walker circulation – with a west–east upper air flow, boosted in the west by air spiralling up from cyclonic storms, then subsiding over the eastern Pacific and returning as low-level easterlies (Figure 4.30). These strong south-easterlies move the waters of the cold Peru current away from the coast, maintaining coastal

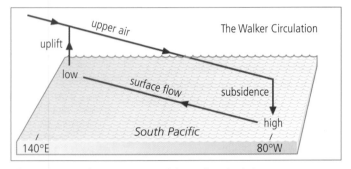

Figure 4.30 *A diagramatic view of the Walker circulation.*

aridity and allowing upwelling water to bring up nutrients that feed a huge fish population.

Dependence on the established atmospheric pattern is emphasised by the serious effects on the Peruvian fisheries, among the world's most productive, when that pattern changes. The normal air stability is illustrated in Figure 4.33, where steam and smoke from a coastal fishery, unable to rise, billows downward – as in Figure 4.40D. Also, as Figure 4.34 shows, many offshore islands are white with guano – droppings from the countless seabirds that thrive on the abundant fish. A periodic change in atmospheric conditions and ocean circulation seriously affects both the environment and the economy.

There are also indications that such disturbances are occurring more often, and, apart from disturbing fisheries, in recent years decreasing guano deposits have pointed to a decline in the seabird population.

Every few years, as the Walker circulation weakens, the ocean current reverses, so that warm surface water flows eastward across the Pacific. Moist air rises offshore, with storms over the Peruvian coastland during a period straddling Christmas – hence known as the El Niño, Christ Child, event. Offshore, warm water now overlies the cold, preventing nutrients rising and greatly reducing fish numbers and dependent bird life.

The cause of this reversal is not fully established, though extra insolation may increase surface water energy and its

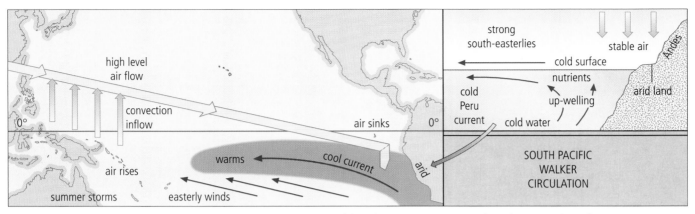

Figure 4.31 *The normal Walker circulation, with easterlies moving waters of the cold Peruvian current away from the coast – the surface water warming as it crosses the south Pacific.* **Inset**: *stable conditions – the coastland remaining dry and nutrients upwelling offshore.*

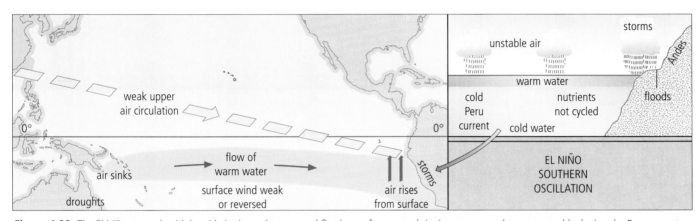

Figure 4.32 *The El Niño reversal, with humid air above the eastward-flowing surface water bringing storms, and warm water blanketing the Peru current.*

atmospheric inputs. Upper air systems are widely disturbed, with weaker Asian monsoons and low rainfall in Indonesia and southern Africa. As sub-tropical subsidence occurs further south, high pressure over north-east Australia brings droughts; while active westerlies cause flooding in California. A strengthened sub-tropical jet stream also affects South

Atlantic conditions: north-eastern Brazil becoming very dry. During a so-called La Niña period the Walker circulation and associated climatic conditions are re-established, completing the El Niño Southern Oscillation (ENSO) – one of several semi-regular climatic fluctuations, including the North Atlantic Oscillation (NAO), described on page 101.

Figure 4.33 *Air subsidence under normal atmospheric conditions causes steam and smoke from a fish-plant, north of Lima, to billow away near ground level.*

Figure 4.34 *Guano – accumulated droppings from numerous sea-birds – covers islands off the Peruvian coast. Much disappears with El Niño rains; at a time when bird numbers are reduced.*

Figure 4.35 *Inversion conditions above Rio de Janeiro, blanketed by mist thick with urban pollutants.*

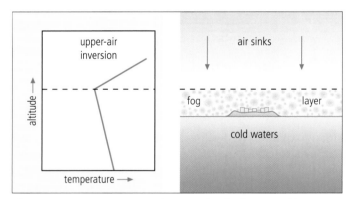

Figure 4.36 *Upper air inversion forms a ceiling for fog over cold water entering San Francisco Bay.*

ATMOSPHERIC CONTAMINATION

Natural and urban-industrial air pollutants may have relatively local effects or be widely distributed by global air circulation. Some enter the stratosphere, especially via vertical movements at fronts, and affect the ozone concentration at that altitude (p. 94).

Urban-industrial pollutants

Power stations burning coal and heavy fuel oils emit many gaseous pollutants, among them large volumes of sulphur dioxide that is responsible for acid rain. This can also harm our respiratory system, as can compounds produced by partial combustion of fuels in vehicles, such as nitrogen oxides, benzopyrene and peroxyacetyl nitrate (PAN), which

Figure 4.37 *Alcatraz island in San Francisco Bay.*

Figure 4.38 *Upper air inversion creates a 'lid' to fog formed over cold water, Alcatraz is barely visible.*

Figure 4.39 *Stable air below low layer cloud over the Blue Mts in New South Wales, causing smoke from Port Kembla's factories and metal-works, rising while hot, to sink and drift fumes among industries and households.*

may also stimulate cancers and damage plant tissues. Ozone, like PAN, is a secondary pollutant. It is not emitted directly, but is formed when strong solar energy acts on combustion products, especially when the air temperature is over 20°C. It, too, can harm the lungs.

Regions like the Mediterranean and southern California experience dry summer climates, sinking air and a high intensity of short-wave energy. In large cities with a high vehicle density, such as Athens and Los Angeles, this combination can lead to photochemical smog. This is rich in dangerous secondary pollutants, as well as very fine suspended asthma-inducing particles from diesel emissions. Other large cities, such as Mexico City and Bangkok, have severe air pollution, with high levels of lead from petrol combustion.

It is technically possible, if costly, to produce and modify internal-combustion engines to eliminate, or greatly reduce, such pollutants. It is thus at least debatable whether some of the incessant publicity and funds earmarked to counter global warming should be transferred to methods of combating this seriously increasing health risk.

Air stability and pollutant dispersal

A plume containing pollutants emitted from a high single stack at a relatively high velocity, and temperature above that of the air about it, is initially buoyant, the aim being to prevent pollutants reaching the nearby surface. However, wind strength and the stability or instability of the local air affect the plume pattern. Figure 4.40 shows behaviour under various conditions.

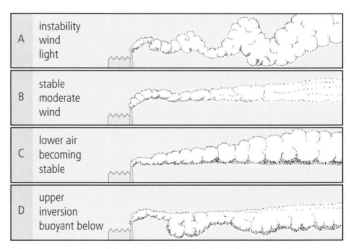

Figure 4.40 *The stability or instability of air has a bearing on the distribution of pollutants emitted from high chimneys.*

A Local instability, as during a hot summer afternoon, creates movements that carry the pollutants up and down in loops of increasing size, and some may pollute the nearby surface.

B Under fairly windy, but stable, conditions the vertical and lateral spread are about equal, so the pollutants fan out into a cone-shaped plume.

C When the lower air becomes stable, as the surface cools in the evening for example, the plume remains elevated and tends to be dispersed upward.

D Under inversion conditions, with warmer air above, the gentle looping movements of the plume have an upper ceiling, bringing pollutants to the surface at intervals.

Figure 4.41 *Sulphurous fumes from copper smelters near Queenstown in western Tasmania created a hillside wasteland. The effects are also seen amid forest on the hills beyond.*

With distance, plumes lose their identity, though may still add to contamination as urban areas with many points of emission act as a single source, extending it downwind for 100 km or more. On a continental scale pollutants from urban-industrial areas may be transported over thousands of kilometres, creating problems in distant countries.

Acid rain

Sulphur dioxide released into the air combines with oxygen and water vapour to form aerosols of sulphuric acid and sulphates, which may eventually return to the surface in acid rain. The combined effects of fumes and acid rain can be devastating, as illustrated in western Tasmania, where sulphur dioxide emissions from copper smelters affected local soils, vegetation and nearby temperate forest (Figure 4.41).

Recently the pH value of rainwater close to industrial areas in western Europe, north-eastern USA and Japan has been below 5.6, compared with near-neutrality (pH 7.0) during the nineteenth century. In parts of Europe and North America pH values have been as low as 2.5, the sulphate ion accounting for over half the acidity. Much comes from factories and power-stations in major industrial regions. Their high chimneys may reduce local pollution, but increase the concentrations over wider areas. It is possible, however, to reduce acidity by fitting de-sulphurisation scrubbers to factories and power-stations. In Britain this has brought considerable reductions in the acidity of emissions.

Plant life is affected when a particular threshold of sulphate ion concentration is exceeded. Conifers and lichens are very sensitive, and in Scandinavia pines have suffered

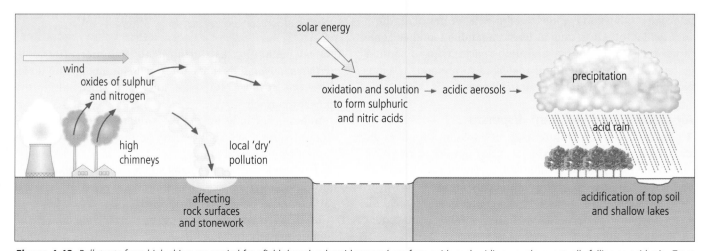

Figure 4.42 *Pollutants from high chimneys carried far afield though a humid atmosphere form acids and acidic aerosols, eventually falling as acid rain. Trees may be directly affected or react to acidified soils.*

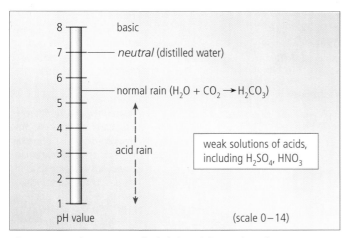

Figure 4.43 *The rain is usually slightly acid, but acidic pollutants can seriously reduce the pH value.*

very badly. There, too, the increasing acidity of lakes over rocks such as granites and quartzites, which tend to produce acid soils, have left many of the smaller ones almost devoid of plant life.

Apart from weak solutions of acids in rainwater, the 'dry deposition' of oxides of sulphur and nitrogen, sinking to the surface, can make soils and lakes more acid. Lakes therefore can increase in acidity by additions of acid rain, by the direct deposition of gaseous compounds, by surface run-off, by infiltration and possibly by delivery through pipes.

Aluminium becomes increasingly soluble as the pH level falls, so that the passage of acid solutions through soils tends to increase the aluminium content of rivers and lakes. Fish can suffer severely if the concentration of aluminium and other toxic metals is increased in this way, and several rivers in north-east England are treating catchment areas to reduce soilwater acidity and prevent the observed decrease in fish

population. Similarly the passage of water with a low pH value through copper and lead pipes can pollute drinking water; lead especially is very much more soluble in acidic water.

Apart from the effects of acidity, increases in other soluble compounds can affect aquatic life-forms, and air-borne pollutants may add to their concentration. The bacteria and blue-green algae in streams and rivers convert atmospheric nitrogen to soluble nitrates, which may be absorbed by plants. An excess of nitrogen may promote algal growth to the extent of depriving other life-forms of oxygen – a process known as eutrophication. As in Figure 4.44 this mostly comes about as nitrogen from cattle excreta passes into solution, and as nitrates from fertilisers wash or seep into the water, for hundreds of millions of tonnes of nitrogenous fertiliser are added to soils each year. But we also release a comparable amount of nitrogen into the atmosphere through the partial combustion of fossil fuels.

QUESTIONS

1 What are the reasons for the following, in which the same chemical reaction is involved?

  a Acid rain tends to create less rural problems in a limestone countryside than in granite terrain.

  b Most buildings faced with limestone show the effects of weathering.

  c Lime is repeatedly added to many of the lakes in Sweden.

2 Even small additions of the oxides of sulphur and nitrogen to the atmosphere can disturb natural chemical and biological processes. Give examples to show how various plants and animals, including ourselves, may be affected.

3 How is air stability, or instability, involved in:

  a creating a blanket of fog containing toxic substances (smog)?

  b the dry deposition of acid-forming gases?

  c the distribution of toxic emissions over a wide area?

Figure 4.44 *Algal growth, stimulated by nitrogeneous wastes, chokes small channels amid these Dutch pastures.*

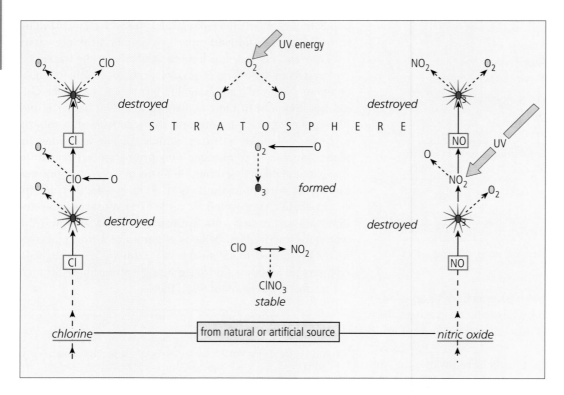

Figure 4.45 *When natural or artificial pollutants release molecules of chlorine and nitric oxide to the lower stratosphere, each may destroy an ozone molecule. Other reactions may then release them to cause further destruction. At the same time, however, molecules of ozone are being created, and some compounds combine to form non-destructive substances.*

Pollutants and stratospheric ozone

The ozone layer – its natural concentration

As short-wave UV solar energy acts on oxygen molecules in the stratosphere, it liberates oxygen atoms and a quantity of heat energy. With nitrogen as a catalyst, an oxygen atom may then combine with an oxygen molecule to form ozone (O_3), whose molecules accumulate in the lower stratosphere, creating the ozone layer.

Over the ages a natural concentration of stratospheric ozone has been established by the chemical actions of pollutants from the biosphere, such as chloromethane and nitrogen compounds from rainforest fungi and those from volcanic outpourings. These and other natural pollutants release elements that destroy a proportion of the stratos-pheric ozone, leaving a 'natural' ozone concentration. This absorbs UV energy, but allows a residue to reach the surface – to which life-forms adapt.

The receipt of artificial pollutants

When chemical pollutants such as CFCs and those from fossil fuels enter the stratosphere, they further lower the ozone concentration. Thus more UV energy may reach the surface, to which life-forms must re-adjust. This poses health problems for humans.

The stratospheric ozone concentration varies with latitude and season. In Equatorial latitudes, where solar energy streams in normally through the year, its peak density is about 22 km up. Above Polar regions solar energy arrives obliquely and only during summer, though throughout the

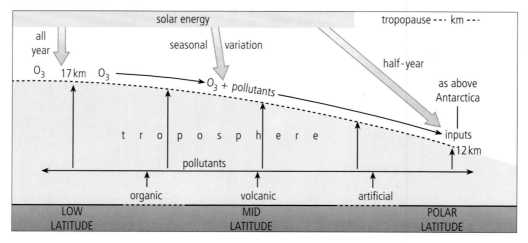

Figure 4.46 *The height of the troposphere varies with latitude, as does ozone production. Pollutants enter the stratosphere, concentrated at certain places, such as frontal zones. With ozone, they can be transported within the lower stratosphere.*

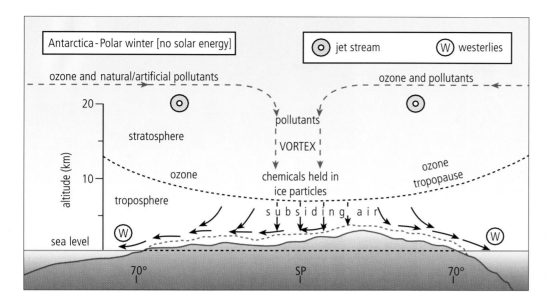

Figure 4.47 *In winter chemical pollutants are carried into the vortex, bounded by strong upper westerlies and, held by ice particles, accumulate in the lower stratosphere.*

day. Here density peaks at a lower altitude and its concentration varies, owing to seasonal changes in stratospheric air currents carrying ozone and pollutants between latitudes (Figure 4.46) and in high-level chemical reactions.

Ozone concentration above Antarctica

During winter air sinks over the cold Antarctic surface. High above, in response to the Coriolis force and absence of friction, the upper air establishes a strong cyclonic circulation (Figure 4.47). Between 20 km and 30 km up air bringing ozone and pollutants from lower latitudes sinks into this polar vortex. There, with temperatures as low as −90°C, minute ice crystals form a high layer that retains the accumulating chemicals.

In spring the sudden rise in temperature melts the crystals, releasing a concentration of pollutants, such as chlorine. The sudden destruction of ozone at that level produces the so-called 'ozone hole' – in fact a considerable fall in concentration. The actual decrease varies from year to year.

As warming continues, the polar vortex breaks down and allows an influx of ozone-rich air from middle latitudes. A sudden rise in concentration is followed by a build-up through summer, when ozone is also created in higher latitudes, mending the so-called 'hole'.

In some springs the fall in concentration may exceed that of the previous year. This occurred successively from 1983 to 1985; but then there was little change until 1988 when the loss was less than in 1987. Other factors influence the ozone concentration, such as the intrusion of particularly destructive volcanic pollutants. After a major eruption they may enter the vortex for several years, depending on the location and volume of emissions. In the late 1970s Mt Erebus erupted within Antarctica (Figure 3.55), and may well have affected depletion during the early 1980s. Following Pinatubo's massive emissions in 1991 there was marked increase in ozone depletion during successive springs. But whatever the effects of natural emissions, it is essential to minimise the release of artificially produced ozone-destructive compounds, such as CFCs.

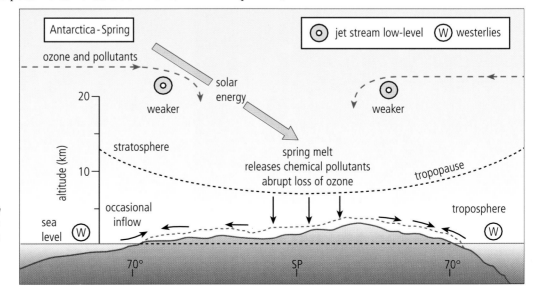

Figure 4.48 *The rapid destruction of stratospheric ozone during spring is followed by an early summer rise in ozone concentration as the vortex weakens and as the solar energy input increases in intensity.*

URBAN GROWTH

There is increasing world-wide development and expansion of large cities, which affect the climate about them, create micro-climates within them and produce and transmit atmospheric pollutants.

Urban micro-climates

A large urban area is more than a collection of buildings; it contains a human population with machinery and vehicles using, generating and releasing heat energy, and discharging wastes. It possesses many different surfaces – roofs, roads, paved areas and parks. High-rise centres, in particular, contrast with more open suburban areas and the countryside beyond, and within them are numerous micro-climates. Their tall buildings and narrow streets form urban canyons, reducing long-wave radiation from surface to the atmosphere and channelling air flows among them.

Air energy and movement within the canyons vary during the day, responding to what is illuminated from sunrise to sunset. Thus in a north–south canyon the upper part of an east-facing wall is irradiated first, with energy from the low sun almost normal to its face. The lower parts, street and west-facing walls remain shaded. As the rest of the wall is gradually lit, the angle of incoming rays to the face decreases. About midday the sun *may* directly heat the street, though the angle of incidence depends on latitude.

Gradually the west walls are illuminated and part of the radiation is reflected back to the buildings opposite, whose materials retain some of their heat. Such exchanges vary considerably in streets of different width and between those running east–west and north–south.

Through the day heat energy dissipated into the air creates turbulence among the buildings, and a proportion is removed from the urban area by convection. At night the heat retained in the materials is released and acts to offset energy emissions, which tends to keep inner cities warmer

Figure 4.50 *Urban micro-climates are not only affected by aspects of streets and avenues, but also by heat and gaseous emissions from traffic and industrial/commercial buildings.*

than their surroundings. However, the overall weather conditions will affect any of these energy exchanges.

The buildings themselves disturb prevailing winds. Figure 4.51 shows how they may influence air flows, cause ground-level pollution and create currents with unpleasant and even dangerous effects. In A air flowing above narrow streets and low housing can cause particles to circulate in eddies between buildings. In B a tall building creates a downwash effect in the lee, so that emissions from low chimneys are taken down to ground level. The concentration of pollutants will depend on whether meteorological conditions favour turbulence or subsidence.

Apart from pollution, strong air currents between buildings and through narrow streets can be unpleasant. Figure 4.52A illustrates that the Venturi effect increasing air speed through a narrowing passage may prove unfortunate if the

Figure 4.49 *Urban canyons in Manhattan show contrasts between sunlit frontages on the west side of the north–south avenues and shaded buildings on the east, with sunshine lighting-up the east–west streets.*

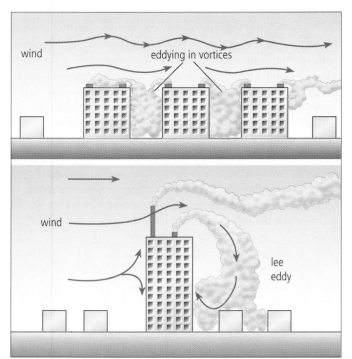

Figure 4.51 *A: Vortices established in the lee of buildings set up persistent eddying, drawing emissions inward and downward. B: Smoke drawn downward in the lee of a tall building sinks among the lower ones.*

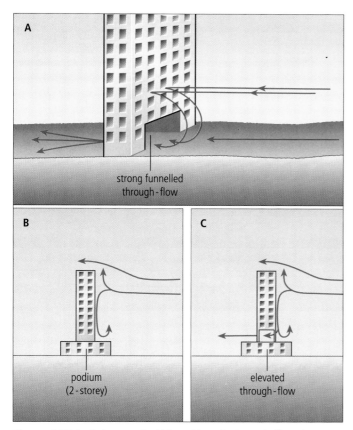

Figure 4.52 In **A** downward eddying adds to the Venturi effect of air accelerating through the ground-level opening. **B** shows the effect of a podium designed to decrease ground-level swirling, and **C** further reduction with a through-flow at higher level.

opening leads, say, to a shopping mall. Swirling downflows at the base of tall buildings, whipping up dust and litter, can also make potential shoppers move on. This effect is sometimes reduced by placing the high-rise building on a podium several stories high, so that down-flows swirl on the podium roof (Figure 4.52B). An elevated throughway above further reduces swirling.

Urban-industrial air pollutants are not simply noxious, they absorb energy and radiate it back to the surface, and can act as condensation nuclei, creating fogs. But controls can be effective, as demonstrated in London, where smoke-free zones established in 1957 have virtually eliminated the notorious 'pea-soup fogs' and smogs thick with soot and sulphur dioxide. Trees in parks and along streets reduce urban noise and help to absorb and break-down gaseous pollutants – London's plane trees being particularly effective.

Of course the actual climatic modifications caused by urban conditions vary from city to city. Counteraction deemed beneficial for some may be unsuitable or unnecessary for others, and obviously the influence of buildings, narrow streets, or open spaces will vary between cities in cold climates and those in the tropics.

Figure 4.53 In Toronto's city centre this tall white building stands on a two-storey podium, while entrances to the black buildings are shielded by wide canopies. Frontages of smaller high-rise buildings are filled-in or screened. Beyond, clusters of tall buildings are spaced out through the suburbs, partly to avoid airflow problems and partly to prevent commercial concentration.

QUESTIONS

1 A city is said to make its own climate. Explain why in fact it creates a collection of micro-climates.
2 Which urban climatic conditions might concern a shop-owner in a city centre, and why?

Figure 4.54 An urban park forms a microclimatic enclave where snow lies deep after it has disappeared from nearby roads. In summer solar energy exchanges with the trees and grass differ greatly from those with the adjacent, often shimmering, street surfaces.

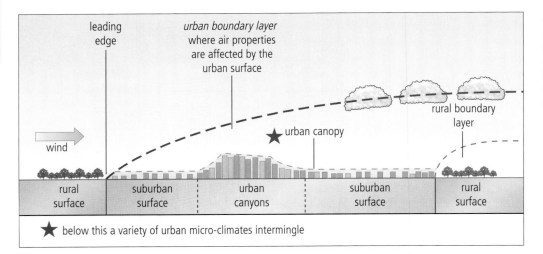

Figure 4.55 The atmospheric layering produced by urbanisation, with the lower urban canopy enclosing numerous micro-climatic features. There is a tendency for urban pollutants to trigger cloud formation downwind.

URBAN-RURAL CLIMATIC CONTRASTS

Wind speed and turbulence

As wind blowing across the countryside encounters the edge of a large urban area, it usually passes successively over uneven clusters of buildings with open spaces, through closely built suburbs and over the high-rise centre, with a similar sequence beyond. Such obstacles affect the wind direction and speed and, as frictional drag causes turbulence, their effects extend up through the overlying air to a height where the general air flow continues unchecked – the **urban boundary layer** (Figure 4.55).

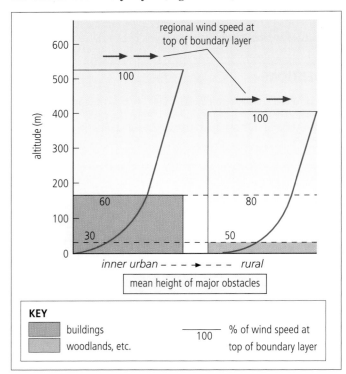

Figure 4.56 The proportion of wind speed to that of free flow varies with height above urban obstacles. There is a relatively freer flow over the more open countryside.

Figure 4.57 Open countryside in East Anglia. In contrast to rapid urban run-off, a large proportion of storm water enters the soil and is slowly discharged into channels such as that in the foreground. However, the surface is open to the effects of drying low-level winds.

Figure 4.56 shows wind speed at the top of this boundary layer and the way it decreases towards the surface. Below roof-level the air is not as calm as the graph indicates, and between a complex inter-mingling of buildings there are often strong winds.

Apart from wind speeds, there are almost always considerable rural-urban contrasts in air temperature and precipitation. Figure 4.58 compares the responses to rainstorms in the countryside with those in an area of suburban

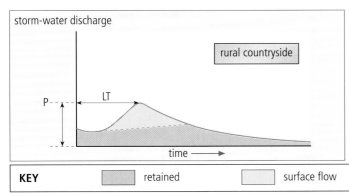
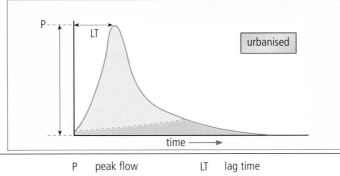

Figure 4.58 *Rapid run-off as a result of urbanisation is indicated by the large peak discharge and the short time to reach the peak, in contrast to the delayed discharge in the countryside, whose soils allow greater infiltration.*

Figure 4.59 *After a storm the surface water channelled from road surfaces collected on a path behind recently developed housing in Milton Keynes, below a firm grassy surface.*

development, where a high proportion of storm rainfall becomes surface run-off and control by channels and pipes becomes essential. Natural vegetation or meadowland provides an intercepting cover and the soil allows infiltration, so that much of the storm water is held until saturation finally allows a delayed surface flow. But as the suburbs spread and houses, formal gardens and streets replace the natural vegetation or cultivated land, surface run-off after a storm is much more rapid, with the danger of streams overflowing. Constructing artificial banks and storm-sewers can check local flooding, but may increase the flood hazard downstream.

Urban heat islands

Incoming solar energy is temporarily 'stored' in buildings and other urban materials, the buildings themselves are warmed by people's activities and heating devices. Also, energy may be transmitted from building to building. Air pollutants about the city absorb some solar energy and their particles act as nuclei for condensation in moist air and so tend to increase cloud cover, which returns a proportion of the energy emitted from the surface. London tends to be cloudier than surrounding rural areas.

Urban areas also lose less heat through evapo-transpiration than the countryside, so that overall these causes combine to make the urban canopy layer a 'heat island', whose intensity is usually greatest a few hours after sunset and least in the middle of the day. After sunset urban areas hold heat longer than rural surroundings, which lose long-wave energy more directly and have moister surfaces. The contrasts between the two are greatest when the sky is clear and the air is still.

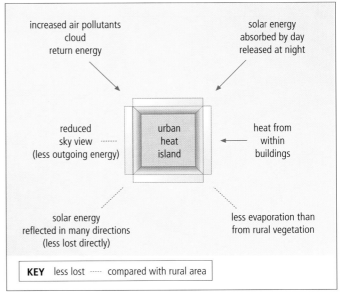

Figure 4.60 *The retention of heat received by an urban area increases, and for several reasons losses are less than in the rural surroundings.*

Greater London as a heat island

Figure 4.61 shows a heat island over London under calm conditions. Mean annual figures point to significant contrasts with the surroundings, though on a mild, windy day any differences would probably be slight. The minimum temperatures show a typically steep rural-urban gradient, with a 'tableland' of warm air about the central part of the city. There will obviously be variations throughout caused by such features as parks, water surfaces, denser and less dense housing, and industrial concentrations.

Topography and people's responses to it affect the pattern. Warmer conditions extend northward along the Lea valley, with its housing developments, whereas there is a cooler wedge in more open country to the north-west. Eastward along the Thames cool air over low-lying estuarine land shows up between warmer air amid denser housing to the north and south.

Regional wind speed and direction affect both the intensity of the heat island and the air temperature distribution. Light westerlies blowing along the line of the Thames valley tend to displace higher temperatures eastward, with a sudden decline further east.

Urban pollutants are mostly carried eastward where, as nuclei of condensation, they tend to increase precipitation downwind of the city centre – though summer thunderstorms usually become more intense over the city area. In winter snowfall is apt to be lighter than over the countryside and may turn to sleet in central London.

Figure 4.62 In Toronto the pattern of heat exchanges and urban warming is complex, with its variations from high-rise centre, to urban nuclei, to open country. The lake also modifies the climate. Its cold winters amplify urban-rural contrasts, but the hot summers mask them.

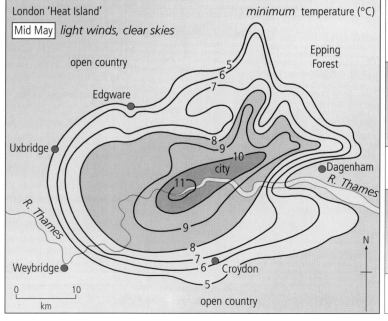

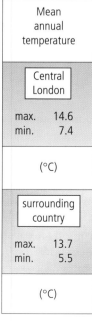

Mean annual temperature		
Central London		
max.	14.6	
min.	7.4	
(°C)		
surrounding country		
max.	13.7	
min.	5.5	
(°C)		

Figure 4.61 London develops as a 'heat island' when open skies and little air movement make for contrasts in minimum temperature between various parts of the city and with the open countryside (after Chandler, 1962).

Urban effects on global warming – points to consider

It must again be emphasised that the existence of such a heat island depends on urban location in a suitable climatic region, and that conditions in many parts of the world will act against rural-urban differences as described above. Nevertheless, wherever the city, micro-climatic responses to conditions of shelter or exposure, manufacturing or farming, are likely to differentiate certain urban characteristics from those of the adjoining countryside.

The direct emissions of heat energy from a large urban area are proportionally very small compared with other energy inputs into the global atmosphere. However, the growing number of urban-industrial areas are emitting more and more gaseous and particle pollutants capable of increasing the local air temperature. They are also transmitting them world-wide.

A quite different consideration is that meteorological stations provide past and present statistics for computer modelling used to assess the extent of global warming. Yet many stations are located within, or close to, urban areas, so that the temperatures recorded and provided may have been influenced by urban heat effects. Such statistics over a given period may thus be unreliable if taken to indicate a trend towards warming on a global scale, unless informed adjustment is possible.

SUMMARY

This chapter emphasises that weather and climate continuously change, owing to numerous natural causes and to increasing populations contributing energy and pollutants to the atmosphere. Semi-regular oscillating changes suggest responses to natural causes, as in the El Niño cycles. In the case of the North Atlantic Oscillation (NAO), when strong sub-tropical Azores high pressure coincides with low pressures about Iceland, the steep gradient makes for westerly air flows with frequent depressions. But when the Azores high is weaker, pressure about Iceland tends to be above normal and blocking highs occur over the Atlantic. These contrasting circumstances alternate in semi-regular decade-long cycles, with corresponding effects on British weather – from wet, windy, mild years to periods with dry, cold winters and hot spells in summer. Over the past few decades the low pressure/warm winter phases have been more frequent.

We can undoubtedly affect local weather conditions, though the extent to which our inputs may affect global climates is uncertain. Chapter 5 considers this and also what our attitude should be in view of such uncertainty.

The effects of atmospheric pollutants on the environment and the fact that they can be a health hazard have also been stressed. It should be a matter for concern as to whether our purpose in curbing emissions should be primarily to clean-up our local environment or to prevent an increase in global temperature. These aims are of course inter-related, and action in one field may help in another. However, making one or the other the main priority can affect how political pressures should be directed world-wide.

QUESTIONS

1 Point to examples of threats to local environment and discuss whether or not these should perhaps concern us more than changes in global atmospheric conditions.

2 Give reasons why Antarctica can be considered as a controlling influence on global climates. Explain why, on the other hand, the effects of global atmospheric changes on Antarctica could be responsible for environmental disturbances world-wide.

3 Give examples to show why the nature of the stratospheric ozone layer depends on the balance of chemical reactions at that altitude. How in the past did this layer maintain a relatively stable ozone concentration?

BIBLIOGRAPHY AND RECOMMENDED READING

Allen, H., 1998, *Quaternary environmental change, Geography*, 83(2), 163

Binns, T., 1990, *Is desertification a myth?, Geography*, 75(2), 106

Brammer, H., 1990, *Floods in Bangladesh, Geographical Journal*, 156(1), 12 and 156(2), 158

Brammer, H., 1996, *Can Bangladesh be protected from floods?, Geography Review*, 9(4), 21

Chambers, F., 1998, *Global warming, Geography*, 83(3), 266

Chandler, T., 1962, *London's urban climate, Geographical Journal*, 128(2), 79

Goudie, A., 1992, *Global warming and the arid lands, Geography Review*, 6(1), 37

Henderson-Sellers, A. and Robinson, P., 1999, *Contemporary Climatology*, Longman

Kirby, C., 1991, *Urban air pollution, Geography*, 80(4), 375

Littmann, T., 1991, *Rainfall, temperature and dust storm anomalies in the Sahel, Geographical Journal*, 157(2), 136

Middleton, N., 1998, *Acid rain in Norway, Geography Review*, 11(4), 6

Rycroft, M., 1990, *The Antarctic atmosphere, Geographical Journal*, 156(1), 1

Shahgedova, M. and Burt, T., 1998, *Urban heat islands, Geography Review*, 11(3), 36

Washington, R. and Palmer, M., 1999, *The North Atlantic Oscillation, Geography Review*, 13(2), 2

Warrick, R., 1988, *Carbon dioxide, climatic change, & agriculture, Geographical Journal*, 154(2), 221

Whyatt, D. and Metcalfe, S., 1995, *Sulphur emissions in acid rain, Geography Review*, 80(4), 375

Wright, D., 1998, *Greenhouse effect, or not?, Geography Review*, 12(1), 10

WEB SITES

Carbon Dioxide Information Analysis Center USA –
http://cdiac.esd.arnl.gov

CSIRO: The greenhouse effect –
ttp://www.dar.csiro.au/pub/info/greenhouse.html

Intergovernmental Panel on Climatic Change –
http://www.ipcc.ch/

NASA's El Niño's site –
http://nsipp.gsfc.nasa.gov/enso/

The UN Framework Convention on Climatic Change –
http://www.unfccc.de/

Chapter 5
Changing climate – debatable issues

Chapter 4 has shown that there are many reasons why climates continue to change at all scales, locally and globally. How such changes are likely to affect present and future generations and their environment and the extent to which we should try to control climatic change are debatable, controversial issues. Remember that we are considering complex integrated systems – solar, atmosphere, land and oceans. No absolutely acceptable reason for a global climatic change has been put forward, so that here it is not a question of 'right' or 'wrong', but of carefully establishing the relevant facts affecting each of the following issues. The pros and cons should be carefully considered and reasons for personal opinions clearly expressed. In setting this out the following section should help.

Skills: Presenting relevant points
- Be as precise as possible.
- Select the most relevant points – if possible, in order of priority.
- Carefully consider the scale involved – local, global, over what period?
- Look for possible trends – are they likely to continue, or perhaps reverse?
- Is the trend part of a cycle – one that may have occurred previously?
- Chapter 4 in particular gives examples of change and probable causes, but look for others from the daily media or library. Some may seem valid, others just 'hype' – balance one against the other.
- Be aware of exactly what statistics express, especially 'mean figures' (see Inset 2.1).
- Consider relevant statistics from several sources if possible. Look at the way they have been obtained and how they are presented.
- Finally summarise the pros and cons and express an opinion – which could just be 'it is better to ...'.

CAN PRESENT WEATHER CONDITIONS AND TRENDS INDICATE CLIMATIC CHANGE?

A period of unusual warmth, exceptional rainfall, or severe storms is apt to be regarded as evidence of climatic change in a particular direction. Thus the popular press responded to the very mild winter and early summer heat of 1992 with 'Greenhouse effect already underway and worsening' and 'Trees and crops will vanish as Earth warms'! Yet in 1974, in response to downward temperature trends during the 1960s and 1970s, the book accompanying a 2-hour TV programme had the following introduction: 'Lost harvests in many countries in the 1970s speak of a turn for the worst in British climates ... The threat of a new ice age turns out to be more ominous than experts thought, even a few years ago.' In debating what such contrasting climatic periods may indicate in the long term, it is worth considering that the hot, dry summers of 1772–83 were followed by exceptionally cool ones between 1809–19.

Predictions of increasing climatic instability invariably follow exceptional storms, as that which swept across southern England in October 1987. Yet in July 1738 a destructive hurricane showered Hertfordshire with ice lumps over 20 cm across, and the one on Derby Day 1911 killed 17 people and 4 horses.

On the global scale one should consider whether apparent climatic abnormalities in one part of the world are part of a global trend or not. Certainly abnormal weather conditions occur simultaneously in countries far apart without any apparent relationship. For example, in May–June 1998 northern India experienced severe heat and drought, but Britain had exceptionally mild, wet weather, with flooding. On the other hand, the El Niño event is certainly accompanied by disturbed weather across the globe. But El Niños occur at intervals – though their disturbing effects may now be more frequent and more intense.

- Above all, when considering this issue, acknowledge the fundamental difference between 'weather' and 'climate'.

THE RATE OF CHANGE – THE RELIABILITY OF STATISTICAL FORECASTS?

Over the last hundred years global mean surface air temperature has increased by between 0.3C° and 0.6C°. In 1995 the International Panel for Climatic Change predicted a further rise of 0.5C° by 2050. Also in 1995 a Meteorological Office computer model, allowing for increasing inputs of pollutants, predicted a mean temperature increase of 0.2C° per decade, and thus a rise of 1.2C° by 2050.

In 1997 the US National Oceanic and Atmospheric Administration, after satellite monitoring of temperatures in two hemispheres over 18 years, recorded that 'Statistics based on satellite monitoring show a much smaller warming trend since 1979. They point to a global cooling of 0.2°C per century.' Again, scale and availability of sufficient information are important. In August 1999 a broadcast weather summary said that Edinburgh had that day experienced the hottest temperature ever recorded there – but failed to say that the station providing the fact had recorded statistics for only 50 years or so, whereas nearby stations provided instances of four or five higher temperatures in Edinburgh since the mid-nineteenth century.

- In view of the difficulties in obtaining sufficient data for large stretches of ocean or for arid areas with few inhabitants, consider the value of obtaining more widespread information via new technology – from satellites, for instance.
- When assessing long-term trends, why may it be misleading that many statistics provided over the last century or so have been from stations within, or adjacent to, large urban areas?

ENERGY INPUTS WITH CYCLIC VARIATIONS – THEIR INFLUENCE?

It is certain that variations in the amounts of solar energy received will continue, owing to regular changes in Earth's rotation and inclination as it orbits the sun and, as satellites now show, to continuous variations in output of energy from the sun itself. The latter are as yet unpredictable, but the occurrence of sunspots and their energy-emitting faculae appear to have a cyclic pattern to which global climates respond. At present it is not fully understood how this occurs.

- What is debatable is the extent to which our activities are modifying the effects of such natural cycles and inputs of energy. Consider changes that took place when a small human population could have had little effect, and whether changes today have similar causes, but are perhaps accelerated by human inputs.

TO WHAT EXTENT CAN NATURAL PHYSICAL PROCESSES AFFECT THE CLIMATE?

Over a long term, plate movements affect the climate of a migrating land area. There are climatic responses to warming following the last Ice Age, varying with latitude and altitude. Over a short term, volcanic emissions have considerable climatic effects. Apart from major eruptions, there are continuous emissions from numerous active sources worldwide. Oceans also have a variety of climatic relationships. Their temperature affects carbon dioxide solubility and influences the distribution of marine life, in particular the

immense number of phytoplankton, which exchange carbon dioxide and other gases with the air above. The climatic effects of ocean currents involve a 'chicken and egg' situation – their location and movements closely respond to air temperature, pressure and winds, yet they themselves influence each of these climatic elements. A change in the pattern of ocean currents can alter air temperatures and rainfall amounts in various latitudes, affecting the climate of adjoining coastal regions.

- Again one should consider the scale at which natural processes may effect climatic change. Some local responses can have large-scale repercussions. Present melting of Arctic glaciers injects cold sinking waters into deep ocean circulation. Increasing amounts *might* cause the North Atlantic Drift to shift southward. Consider how this would affect Britain's climate.

HUMAN ACTIVITIES CAN AFFECT THE CLIMATE – BUT ON WHAT SCALE?

As indicated, urbanisation creates local micro-climates and under certain meteorological conditions affects the regional climate. While heat energy from a single urban source has little direct effect on atmospheric warming, urban gaseous and particle emissions can be widely distributed, creating environmental problems. Replacing natural vegetation with pasture or cultivation disturbs surface-atmospheric exchanges, surface temperatures and wind strength, often with considerable effect on the local climate.

In such cases it is essential to present events on the correct scale. Unfortunately on the global scale many environmentalists exaggerate the extent and climatic effects of disturbance to create awareness of potential problems.

Figure 5.1 Disturbance effects micro-climatic changes. A barchan dune amid the northern Tanzanian savanna, where overgrazing and trampling created a patchwork of tussocks and bare surfaces, causing microclimatic changes. In the dry season the high surface temperature of bare patches, with a large lapse rate above, creates mini-turbulence and raises dust. Wind moving freely over the tableland has created barchan dunes.

Figure 5.2 Warming can bring advantages. In Canada, as rising temperature caused ice to retreat, outwash over a landscape of kettle holes and small lakes spread sufficient fertile moraine to support today's extensive farming on this northern fringe of the Albertan prairies. If, by 2050, as some predict, summer temperatures in the Canadian Arctic have risen by 0.5C° and are much higher during winter, growing conditions at Yellowknife would resemble those at Calgary today, a thousand kilometres to the south.

In 1991 one 'green' organisation, rightly concerned with climatic and social consequences of the clearance of rainforest in south-west Brazil to re-settle families from over-populated coastal regions, pointed out that one million acres of rainforest had been lost in the Amazon Basin in 1988 and that 'at that rate within the next ten years much of the remaining Amazon forest could be wiped out'. As its forests then covered some 1200 million acres, their warning erred by over a thousand years! As always, when considering environmental hazards and climatic change, scale matters.

- It is easy to account for local climatic effects of surface disturbance, as in Figure 5.1, but more difficult to assess effects on the global scale. Forest clearance can cause drastic surface erosion and increase atmospheric carbon dioxide, yet the effects vary with natural regeneration and with conversion to agricultural systems and managed plantations.

POSITIVE CONSEQUENCES OF CLIMATIC CHANGE?
So often the response to an indication of climatic change is to point to the negative consequences. This suggests that over Earth's surface the present climate is ideal and that any variation in temperature, humidity or precipitation must be undesirable, with action required to prevent it. In fact global warming in itself and accompanying changes in, say, precipitation may benefit the population in many parts of the world. Figure 5.2 shows how climatic change has freed fertile agricultural land. In many places the disappearance of low-level ice has allowed settlement on fertile soils. Each region needs to be considered separately. Sea-level rise can also have positive effects as well as negative ones.

- Suggest where warmer, wetter conditions might well prove beneficial, and give other reasons why responses to the effects of global warming might be welcomed.
- One point to debate is whether we should act to prevent global warming, or establish projects adapted to deal with the likely effects of a warmer climate.

MINIMISING CLIMATIC CHANGE – THE NEED FOR ACTION?
Another consideration is that as there are uncertainties about the consequences of climatic change, should there be action to minimise such change? Certainly an increase of global mean surface air temperature by 1C° over half a century would cause changes in plant associations, crop distributions, wildlife habitats and sea-level; while more variable weather conditions could lead to further droughts or floods in particular areas – hence the need for action. Developed countries should be able to cope with the effects of a slowly changing climate, but the less developed, with rapidly increasing population, may be less able to do so. The debate here is between two attitudes – 'Why take chances? Act to minimise climatic change' and 'Why waste resources on "possibilities"? There are certainly other environmental problems needing attention and funding.'

- Once again scale is involved. As developing countries with large populations aim to be on a par with industrially developed nations, they will increase their number of fuel-powered vehicles and establish numerous industries with gaseous emissions. Thus the possibilities of pollutant-induced climatic change must increase. If global warming is considered likely, and seen as undesirable, the alternatives would be to induce them to hold back such development, or provide international co-operation to minimise noxious outputs.

MINIMISING CLIMATIC CHANGE – CAN CURRENT PROJECTS PROVE EFFECTIVE?

Projects exist and measures are being taken to cut back emissions of carbon dioxide from artificial sources. Whether this is essential or merits the proposed expense is another subject for debate, as is the extent to which reduction can be achieved. Will alternative methods of power generation, avoiding the use of fossil fuels, be sufficient and effective – especially in view of increasing energy demands by a rising population aiming at higher living standards? Might not practical methods of preventing emissions of noxious gases from fossil fuels, on health grounds, be a more suitable target for expenditure – even allowing carbon dioxide to contribute to a temperature rise, if need be? These are certainly debatable issues.

- What would be the advantages of turning to other means of generating energy as alternatives to burning fossil fuel and its combustion in vehicles? Remember that in using electricity to power vehicles there must be an acceptable way of generating such energy on a large scale.
- Would the main objective of such alternatives be mainly to reduce global warming arising from carbon dioxide and other gases, or to create a cleaner, healthier lower atmosphere?
- Some energy sources have patent disadvantages. Consider whether the following drawbacks should be regarded as sufficient to prevent their introduction as alternatives to the use of fossil fuels.
  a The use of wind-powered generators involves erecting clusters of large structures in the open countryside – or offshore.
  b Dams large enough to generate and distribute hydro-electricity over a wide area may involve displacing populations, damaging historic sites, being a local hazard through possible damage by earth movements, and may become less effective as silting reduces storage capacity.
  c Using nuclear energy involves risk from radioactive emissions through damage to the plant by natural causes, or human error, and through disposal of nuclear wastes.
  These and issues concerning other energy sources, such as solar energy, may suggest that it is preferable to continue to use fossil fuels, despite their declining reserves, and to concentrate on ensuring cleaner emissions after combustion.

Skills: Being precise

It is extremely difficult to be realistic about the behaviour of Earth's atmosphere, so that in dealing with any aspect of weather or climate it is important to identify the relevant factors and to be as precise as possible. The following are some of the essentials to be considered.

1 **The scale in space and time** – of a storm; of the period of heat or cold; of flood or drought.
2 **The frequency at which an event occurs** – whether a 'one off' or perhaps a cyclic occurrence; whether it is happening more frequently/less frequently.
3 **Relationship to environmental conditions** – cloud behaviour indicating instability and possible consequences; prevailing influences of systems such as cyclones or anticyclones.
4 **Wider influences** – volcanic emissions; the behaviour of ocean currents.

Examining precisely the factors affecting a weather situation or climatic condition avoids the danger of simplification. Certain over-simplified, eye-catching headlines and buzzwords are used by the media, by political commentators and by the public in order to portray them as environmental threats, without any revelation of their actual characteristics. Among these are 'greenhouse effect', 'ozone hole', 'desertification', and 'global warming' itself. In reality, as we have seen, the greenhouse effect is a vital process for all life-forms, and the mis-use of the terms 'hole' and 'desertification' have been stressed above. Beware of buzzwords!

Appendix

CLIMATIC CLASSIFICATION – TREWARTHA'S MODIFICATION OF THE KÖPPEN SYSTEM

Group A

TROPICAL RAINY CLIMATES. Temperature of the coolest month over 18°C

Af *No dry season.* Driest month has over 60 mm. Inter-Tropical Convergence Zone, with mT (or mE) air masses.

Am *Short dry season,* but rainfall sufficient to support rainforest (wet monsoon type).

Aw *Dry during the period of low sun (winter).* Driest month under 60 mm. Dry cT air in winter. *Wet during period of high sun,* when ITCZ moves poleward and moist mT air flows in.

Group B

DRY CLIMATES. Evaporation exceeds precipitation. W – arid; desert. S – semi-arid; steppe. Boundaries between *BW/BS* identified by formulae based on annual rainfall and mean annual temperature.

BWh *Hot desert.* Mean annual temperature over 18°C. Source regions of cT air: sub-tropical Highs. Dry cT Trade Winds.

BSh *Tropical and sub-tropical semi-arid.* A short rainy season – cT air for most of the year.

BWk *Middle latitude interior desert.* cT air masses in summer and cP air masses in winter. Large annual temperature range. Persistently dry.

BSk *Middle latitude semi-arid.* Dominated by dry cP air in winter; mainly cT air in summer; meagre rainfall, mostly in summer.

n (nebel) is used to show frequent fogs along coastlands with cool waters offshore.

Group C

HUMID MESOTHERMAL (Moist Temperate). Temperature of coldest month between 18°C and 0°C.

Cs *Sub-tropical, dry summers.* At least three times as much rain in wettest month as in driest summer month. Driest month less than 30 mm. Summers dominated by sub-tropical High (the *Cs* regions lie on the stable eastern side of the Highs). mP air in winter, with cyclonic storms and rain.

Csa *Hot summers.* Warmest month averages over 22°C.

Csb *Warm summers.* Warmest month averages under 22°C.

Ca *Humid sub-tropical, hot summers.* Warmest month over 22°C. In summer moist mT air from the unstable western side of the sub-tropical High over ocean. Winters with cP air invading, and cyclic storms developing.

Caf *No dry season.* Driest month over 30 mm.

Caw *Dry winters.* At least ten times the rain in the wettest summer month as in the driest winter month.

Cb *Marine climate, cool-warm summers.* Warmest month under 22°C. Mostly middle latitude west coasts which receive moist mP air and series of depressions. Rain in all seasons.

Cbf *No dry season.* The most common *Cb* type (*Cbw* describes parts of south-east Africa).

Cc *Marine climate, short cool summers.* Warmest month below 22°C; less than four months over 10°C, rain in all seasons.

Group D

HUMID MICROTHERMAL (Rainy/Snowy, Cold). Temperature of coldest month under 0°C, and warmest month over 10°C.

Da *Humid continental, warm summers.* Warmest month over 22°C. Precipitation in all seasons, accent on summer maximum; winter snow cover. Zone of frequent clashes between polar air and tropical air. Variable weather.

Db *Humid continental, cool summers.* Warmest month under 22°C. As for Da, but long winter snow cover.

Dc *Sub-Arctic.* Warmest month below 22°C; less than four months over 10°C; winter cP air mass; cold stable air; summer occasional cyclonic storms with mP air; precipitation light; low winter evaporation, so remains moist.

Dd *Sub-Arctic, with very cold winters.* Coldest month below −38°C. Very light precipitation.

Group E

POLAR. Temperature of warmest month less than 10°C.

ET *Tundra.* Warmest month above 0°C. mP, cP and A air masses interact; cyclonic storms, light precipitation, mainly in summer.

EF *Ice-cap, perpetual frost.* No month over 0°C mean temperature; source regions for Arctic/Antarctic air masses.

H is also used to identify highland areas with climatic conditions varying with altitude.

Abbreviations

Air mass characteristics – cP; mP; cT; mT; cA; mE (P Polar; T Tropical; A Arctic; E Equatorial; c continental; m maritime)

CFCs chlorofluorocarbons – compounds used in refrigerants and spray-can propellants, whose photochemical break-up releases free chlorine, capable of destroying stratospheric ozone.

Climatic Regions *A–E* (with secondary and tertiary symbols) see p. 48.

DALR Dry Adiabatic Lapse Rate – that in rising air whose water vapour is not condensing.

Ea the actual loss of water from soil and plant cover by evapo-transpiration.

ELR Environmental Lapse Rate – that in the air about rising air under observation.

Ep the potential amount of loss by evapo-transpiration that would leave sufficient water to maintain the vegetation cover – can be calculated.

FCD a Flood Control and Damage Limitation project related to inundation in Bangladesh.

ITCZ the Inter-Tropical Convergence Zone – a broad zone of relatively low pressure between the northern and southern Trade Winds. It tends to move poleward with the 'overhead sun'.

PAN peroxyacetylnitrate – a secondary atmospheric pollutant created by solar energy acting on combustion products from vehicles.

SALR Saturated Adiabatic Lapse Rate – that in moist air receiving latent heat from condensation.

UV ultra-violet radiant energy of wavelengths slightly shorter than 0.4 μm.

Units

1°C	the actual temperature of one degree Celsius (°C = °K – 273°)
1C°	an increase or decrease of one degree Celsius
1 centimetre (cm)	0.394 inches
1 kilometre (km)	0.62 miles
1 knot	11.5 mph
μ (micro)	one millionth

TEMPERATURE CONVERSION

°C	–70	–60	–50	–40	–30	–20	–15	–10	–5	0	1	2	3	4	5	6	7	8	9	10
°F	–94	–76	–58	–40	–22	–4	5	14	23	32	34	36	37	39	41	43	45	46	48	50

°C	11	12	13	14	15	16	17	18	19	20	21	22	23	24	25	26	27	28	29	30
°F	52	54	55	57	59	61	63	64	66	68	70	72	73	75	77	79	81	82	84	86

°C	31	32	33	34	35	36	37	38	39	40	41	42	43	44	45	46	47	48	49	50
°F	88	90	91	93	95	97	99	100	102	104	106	108	109	111	113	115	117	118	120	122

(°F Equivalent to nearest whole number)

Glossary

Aerosol tiny particles, solid or liquid, suspended in a gas.

Adiabatic process a change in air pressure and volume without gain or loss of heat from the environment.

Advection a horizontal movement of a body of air over Earth's surface.

Air mass a mass of air of broadly homogenous temperature and humidity covering a large source area, bounded by frontal surfaces or transition zones.

Albedo the proportion of radiant energy reflected by a particular surface.

Amplitude (of a wave) the maximum distance from its equilibrium value.

Anabatic Wind air that rises up-slope (*Gk. ana* – up).

Anaerobic in the absence of free oxygen.

Barchan a crescent-shaped dune with horns pointing away from the prevailing wind – a gentle windward slope, a steep leeward one.

Barograph a self-recording barometer, tracing pressure onto a moving drum.

Biochemical chemical action affecting living organisms.

Biomass the weight of organic matter per unit area, or in a given medium (e.g. in a soil).

Blocking high an atmospheric anticyclone stationary or moving only slowly eastward.

Centrifugal force the outward-moving force acting on a body rotating about a central point.

Cirrus high wispy cloud of ice crystals – often elongated by upper winds.

Climograph a graph with two or more climatic variables plotted for each month of the year.

Cold front surface of contact with a cold mass forcing its way beneath a warm air mass.

Continentality being continental, as opposed to oceanic – also the extent to which climate is influenced by distance from the sea.

Convection heat energy transferred within a fluid causing the movement of the fluid itself – as that causing expansion and movement of air.

Coriolis force the 'fictitious' force that relates the actual direction of air movement over the surface to the Earth's rotating grid of latitude and longitude – thus apparently responsible for the formation and direction of cyclonic movements.

Cumulus convection-formed cloud masses, rounded and towering above a flat base.

Cyclonic describing the movement of air rotation about a pressure centre.

Desertification a term used to imply the extension of permanent aridity with barren conditions over an area, often over land peripheral to an existing dessert.

Degradation the wearing and disintegration of a surface by erosive processes, with or without the removal of waste material.

Ecliptic Plane the plane of Earth's orbit about the sun.

Epiphyte a plant attached to other vegetation, but not a parasite.

Evapo-transpiration combined loss of water to the atmosphere from the soil and by transpiration from plants.

Facula a bright area of high temperature on the sun's surface, emitting exceptionally intense flares of energy – often apparent amid a darker sunspot area.

Foraminifera microscopic single-celled marine animals with calcareous or complex organic shells.

Front the surface of contact between two unlike air masses.

Garrigue short-lived, sweet-smelling undershrubs, such as lavender, thyme and broom, able to withstand a hot, dry summer.

Geostrophic flow wind moving towards an area of lower pressure but taking a direction established by Earth's rotation.

Gradient wind air moving towards a centre of low pressure; the greater the pressure difference – the closer the isobars – the higher the wind speed.

Gravity wind air becoming denser than the surrounding air and moving downslope.

Halophytic tolerating a relatively high environmental salt content.

Hydrological relating to water – its state, distribution, or movement.

Hygroscopic having a tendency to absorb water.

Icesheet   a continuous mass of ice of great thickness covering a large area of accumulation, from which it moves outward.

Iceshelf   a plate of floating ice attached to an icesheet.

Insolation   energy emitted by the sun that reaches Earth's surface.

Ion   an atom having gained a negative charge by acquiring electron(-s) or a positive charge by losing electron(-s).

Ionosphere   a deep atmospheric zone above the stratosphere where free ions and electrons are concentrated in layers that reflect radio waves to the surface.

Isobar   a line on a map joining places with the same barometric pressure.

Isotherm   a line on a map joining places of the same air temperature.

Isotope   atoms of the same element differing in atomic weight, but with identical chemical properties.

Jet stream   a band of very strong wind at the core of the upper westerlies, just below the tropopause, a few hundred miles wide and several kilometres deep.

Karst   limestone with features such as sinkholes, wide-jointed pavements, steep-sided gorges and towering residual hills produced by weathering, especially by solution.

Katabatic wind   a down-slope wind (*Gk. Kata* – down).

Kettle hole   a hollow left in thick glacial drift as an isolated ice block melted.

Kinetic energy   the energy of a body by virtue of its motion.

Lag time   the time between the reception of precipitation and the peak flood.

Lapse rate   the rate of decrease of air temperature with altitude, usually expressed in C° per 100 m.

Latent heat   energy transferred when a substance changes state – from solid to liquid, liquid to gas, liquid to solid.

Maquis   a scrub vegetation with thick-leafed shrubs such as laurel, myrtle and rosemary, with some taller evergreens – adapted to withstand hot, dry summers.

Maunder Minimum   a period in the late seventeenth century, during the Little Ice Age, when there were very few sunspots.

Meridional   in a direction from tropical latitudes towards the poles – a meridian being a line of longitude.

Nimbus   cloud from which rain is falling.

Occlusion   being cut off (excluded) – as when warm air is lifted from the surface by cold air advancing beneath it.

Organic   of plants and animals, or compounds derived from them – or may refer to chemical compounds of carbon.

Orographic   concerned with bold surface relief, such as mountains.

Ozone   a gas whose molecules each contain three oxygen atoms (O_3).

Periglacial   cold regions bordering ice-covered lands, with freezing/thawing as dominant processes – often with permafrost, and affected by the action of melt-water.

Permafrost   permanently frozen ground beneath a shallow active layer that thaws only during the summer.

pH   a measurement of the acidity or alkalinity of a solution, expressed as a number between 0 (extreme acidity) and 14.

Photochemical   chemical reaction initiated or accelerated by exposure to light.

Photosynthesis   the formation of carbohydrates from carbon dioxide and water by the action of solar energy on green plants containing chlorophyll.

Phytoplankton   microscopic aquatic plants acquiring solar energy – primary producers in the marine food chain.

Piedmont glacier   an icesheet at the foot of a mountain range, with ice contributed by high glaciers from the range.

Plate movement   slow movement of one of the large rigid segments (plates) of continental or oceanic material forming Earth's outer shell.

Polder   drained land, deltaic or reclaimed from the sea, protected by an embankment.

Potential energy   that waiting to be released – as in a coiled spring, or a wheeled vehicle on a hill-top, or stored in clouds.

Roaring Forties   an area roughly between latitudes 40°S and 50°S where surface westerlies blow strongly across wide ocean stretches – sometimes applied to the winds.

Rossby waves   part of a large-scale longwave motion of the upper westerlies – effecting transfers of heat between latitudes.

Saturated air   air holding the maximum possible water vapour at a given temperature and pressure.

Sheetwash   a sheet of water flowing for a period over a relatively gentle surface slope, and apt to produce 'sheet erosion'.

Smog   fog with a concentration of smoke particles acting as nuclei for condensation – often chemically polluted.

Solar elevation   the angular position of the sun above the horizon.

Solfatara   a volcanic vent emitting sulphurous gases and steam.

Stomata pores on plant leaves regulating the amount of water transpired by the plant – their degree of opening controlled by guard cells.

Stratus a layer cloud of water droplets, or at a high altitude of tiny ice crystals.

Sunspots broad, dark, relatively cooler areas appearing spasmodically on the sun's surface yet associated with adjacent outbursts of solar energy, and apparently moving with the sun's rotation.

Super-cooling when water can be cooled well below 0°C and yet remain liquid.

Synoptic relating to a summary of weather conditions.

Sublimation the conversion of a solid direct to vapour without melting.

Taiga coniferous forest covering large areas of Siberia – the name often used for similar extensive coniferous forest in North America.

Tectonic referring to forces responsible for deforming Earth's crust, and to the resulting features.

Temperature inversion an increase in air temperature with altitude, which may occur near the surface (low-level inversion), or higher air temperature encountered at a particular altitude (upper-air inversion) – in each case a reversal of the normal decrease.

Thermal high high pressure induced as a very cold surface creates air subsidence – 'thermal' referring to the temperature control.

Transhumance the practice of moving flocks or herds seasonally between regions with different climatic conditions to provide better grazing.

Transpiration the removal of water from the interior of a plant through pores – mostly through leaf stomata.

Tropopause the upper limit of the troposphere where the temperature ceases to fall with altitude; higher over equatorial regions than about the poles.

Venturi effect the increase in velocity of a fluid flowing through a constriction.

Warm front the surface of contact where advancing warm air moves up and over a cold air mass.

Weathering physical or chemical processes causing the breakdown of rocks at, or near, the surface.

Zonal wind one moving between the parallels of latitude – east–west/west–east movement.

Index

Location Index

Text © David Money 2000

Original line illustrations © Nelson Thornes 2000

The right of David Money to be identified as author of this work has been asserted by him in accordance with the Copyright, Designs and Patents Act 1988.

All rights reserved. No part of this publication may be reproduced or transmitted in any form or by any means, electronic or mechanical, including photocopy, recording or any information storage and retrieval system, without permission in writing from the publisher or under licence from the Copyright Licensing Agency Limited. Further details of such licences (for reprographic reproduction) may be obtained from the Copyright Licensing Agency Limited, 90 Tottenham Court Road, London W1P 0LP.

First published in 2000 by:

Nelson Thornes
Delta Place
27 Bath Road
CHELTENHAM
GL53 7TH
England

00 01 02 03 04 / 10 9 8 7 6 5 4 3 2 1

A catalogue record for this book is available from the British Library.

ISBN 0 17 444712 4

Printed and bound in Croatia by Zrinski d.d. Cadovec

Page layout and illustration by Pentacor, High Wycombe, Buckinghamshire, UK

Acknowledgements

With thanks to the following for permission to reproduce copyright material in this book:
The Telegraphic Colour Library; figure 2.46
The Meteorological Bureau, Perth (W.A.) for permission to use data for figure 2.39

All other photographs by David Money

Every effort has been made to contact copyright holders. The publishers apologise to anyone whose rights have been inadvertently overlooked, and will be happy to rectify any errors or omissions.